EARTH
A PLANET IN
PERIL

HUMANITY,
THE DESTROYER OR SAVIOR?

FRED GRAHAM-YOOLL. 9:1:18

Earth, a Planet in Peril

Copyright   2018 Fred Graham-Yooll

All rights reserved

First Edition

Independently published

ISBN 9781719827393

Printed in the United States of America

Cover courtesy of Wikipedia and 'Forest Fire Protection Services' and Xorex Service who designs the fire detection software.

CONTENTS

Contents	3
Casabianca – Hemans	4
If – Kipling	5
Dedication	6-7
Alert	8
Summary	9-10
Introduction	11
1. Carbon, our Planet's protector	12-27
2. Our Population	28-32
3. Researching Climate Change	33-42
4. Continental Drift	43-48
5. Wind and Ocean Currents	49-60
6. Ocean Level Changes	61-67
7. Water	68-103
8. Energy	104-121
9. Transitioning to Renewables	122-128
Biography	129

Earth, a Planet in Peril

CASABIANCA

The boy stood on the burning deck,
Whence all but he had fled;
The flame that lit the battle's wreck,
Shone round him o'er the dead.
Yet beautiful and bright he stood,
As born to rule the storm;
A creature of heroic blood,
A proud, though childlike form.

The flames rolled on - he would not go,
Without his father's word;
That father, faint in death below,
His voice no longer heard.
He called aloud – 'say, father, say
If yet my task is done?'
He knew not that the chieftain lay
Unconscious of his son.

'Speak, father!' once again he cried,
'If I may yet be gone!'
And but the booming shots replied,
And fast the flames rolled on.
Upon his brow he felt their breath
And in his waiving hair;
And look'd from lone post of death,
In still yet brave despair.

And shouted but once more aloud,
'My father! Must I stay?'
While o'er him fast, through sail and shroud,
The wreathing fires made way.
They wrapped the ship in splendour wild,
They caught the flag on high,
And streamed the gallant child,
Like banners in the sky.

There came a burst of thunder sound-
The boy- oh! Where was he
Ask of the winds that far around
With fragments strewed the sea!
With mast, and helm, and pennon fair,
That well had borne their part,
But the noblest thing that perished there
Was that young faithful heart.

Felicia D Hemans. 1834

IF

If you can keep your head when all about you
  Are losing theirs and blaming it on you;
If you can trust yourself when all men doubt you,
  But make allowance for their doubting too;
If you can wait and not be tired by waiting,
  Or, being lied about, don't deal in lies,
Or, being hated, don't give way to hating,
  And yet don't look too good, nor talk too wise;

If you can dream—and not make dreams your master;
  If you can think—and not make thoughts your aim;
If you can meet with triumph and disaster
  And treat those two impostors just the same;
If you can bear to hear the truth you've spoken
  Twisted by knaves to make a trap for fools,
Or watch the things you gave your life to broken,
  And stoop and build 'em up with wornout tools;

If you can make one heap of all your winnings
  And risk it on one turn of pitch-and-toss,
And lose, and start again at your beginnings
  And never breathe a word about your loss;
If you can force your heart and nerve and sinew
  To serve your turn long after they are gone,
And so hold on when there is nothing in you
  Except the Will which says to them: "Hold on";

If you can talk with crowds and keep your virtue,
  Or walk with kings—nor lose the common touch;
If neither foes nor loving friends can hurt you;
  If all men count with you, but none too much;
If you can fill the unforgiving minute
  With sixty seconds' worth of distance run—
Yours is the Earth and everything that's in it,
  And—which is more—you'll be a Man, my son!

                      Rudyard Kipling   1910

DEDICATION

I would like to dedicate this book to the thousands of scientists and others who have striven to bring the subject of 'Climate Change' to the attention of the people of our World. These four were all pioneers.

Photo; courtesy of the Royal Institute.

John Tyndal, an Irishman, who became intrigued by the discovery that the Earth had been moving in and out of a series of 'Ice Ages.' His work (1859) identified that some common gases (carbon dioxide and methane) may have played a part in that phenomenon.

Photo; courtesy AIP Emilio Segre Visual Archives.

Svante Arrhenius, a Norwegian, who in 1896 developed the first mathematical model that demonstrated the link between these carbon gases and world temperature changes.

Photo; courtesy US Fish and Wildlife Service.

Rachel Carson, an American, while not involved directly with climate research; her book, 'Silent Spring' published in 1962, set off the alarm bells in my head and throughout the scientific community to the damage we were causing to our environment.

Photo; courtesy Twitter.

In this same vein, my next nominee is Al Gore, another American, who, with his 2006 movie, 'An Inconvenient Truth', focused the World's attention on 'Global Warming'.

Earth, a Planet in Peril

ALERT, ALERT, ALERT.

How does one alert the World about subjects so vital that they will determine humanity's very future?

Compounding this issue is watching the misdirection taking place such that valuable time is being lost to the finding of those solutions.

A second problem is to try and understand why our leaders are letting our Planet be destroyed in front of our very eyes.

As for our cosmologists, why are they wasting time looking for other worlds when our own will soon be gone? Don't they realize that our world may be unique due to the way it was formed? And as for distance and travel time, let's not be complete idiots.

Brainwave; why not write a book, but not just any book, write one with a summary at the front so that readers can learn the whole story in seconds, with the rest of the book explaining the details and the evidence.

As for this writer, he may be somewhat unique being in agriculture over his entire career yet working for the World's largest Oil Company and perhaps knowing a great deal of this story and it's solutions.!

*****

Mission accomplished. Everyone is immediately alerted and knows what needs to be done in seconds. After that, it is up to the reader to decide whether they want to learn more and to delve into the rest of this book. The point being that the Alert has been sounded with the proof at hand.

NOTE: To save time and dispute, only Government and their sourced University/research data has been used throughout this book with NOAA, NASA, EPA, DOA, EAI, UN. World Bank, and Smithsonian data dominating. The search for data could not have been done without the aid of Wikipedia.

*****

SUMMARY

In 2006, the alarm sounded that the burning of fossil-fuel was resulting in a build-up of carbon dioxide (CO_2) in our atmosphere. In turn, it was causing temperatures to rise, and was dubbed 'Global Warming.' Few realized that by pointing the finger at fossil fuel as being the culprit, we were about to be sending ourselves off on a 'Wild Goose Chase' by going after a symptom rather than the problem. Allow me to explain:-

When one talks of increasing the population of major mammals from 1 billion to 11 billion in less than 150 years, something has gone far wrong. So, if the growth rate of humanity might be the problem and we had gone off to see our doctor, the first question asked would have been; what are your symptoms? We can be certain that 'Global Warming' would have been only one, and not at the top of the list. And now you can see why this 'Misdirection' might have been fatal losing over 12 years of research time, not so much that it has been wasted, for it has not, and would have been in dire straits without it, but rather, there should have been massive amounts of research covering many other subjects and areas that were equally important.

So what have our tens of thousands of 'Global Warming' scientists discovered? If one were forced to pick out the best from the millions of things discovered, even the scientists themselves might have picked out the 'Ice-Core' drillers in Greenland and Antarctica who uncovered the fact that our Planet had been moving in and out of 'Ice Ages' at least eight times during the last 800,000 years. The frightening thing though, was that the 'Ice Age' part of each cycle had, on average, lasted for over 60,000 years! Also, each time, when we reached the peak of the warm part of the cycle, they found it triggered a drop in temperature that sent our Planet into the next sixty-thousand plus year 'Ice Age'. Oops; cynically, the powers that be said, "we had better change the name of this problem to 'Climate Change,' for we might be going into another Ice Age". The one blessing in all of this is that the climate researchers have given us a much better insight into how our Planet works and reacts to changing conditions. Unfortunately they also discovered that a continuing surge in human population will be taking place due to the fact that as the World's percentage of child-bearing population increases, so too does the birth rate–and all this despite a reduction in growth in the better off areas of our World. While all this is documented in detail, there is a solution to every problem and that is why this book is so important. Not only can this mess be stopped, it can be reversed, but it has to be done quickly.

As an aside, we may have come up with a reasonable hypothesis that explains why these climate swings occur, but there is a problem for by the time it can be proved we might be into the next 'Ice Age', with most of the World's population dying of hunger, cold, and starvation. Misery on a scale that cries out for action by everyone.

Meanwhile, in large part using the research results from these same scientists, we made an extensive study of what other symptoms we could find, with the many uses of fossil-fuel and its replacement by renewable energy, one of the most obvious. The answer from that work was surprising! It said that regardless of 'Global Warming', it made economic and environmental sense to make that switch as soon as possible leaving about 20% unchanged, given today's technology. (Air travel and peak-load electrical production.)

We next explored Agriculture and where it was heading, if the anticipated population is to be fed. It turned out that irrigation and the availability of fresh water was one of the vital inputs and was already accounting for 70% of worldwide use. With 30% coming from underground water, and that source in many cases not renewable, and coupling that with reduced glacier melt causing rivers to become smaller, it raised the question about future fresh water supply. This time we found the answer in ancient history, for without modern technology, they were forced to come up with simple solutions.

The huge surprise for me was in the carbon cycle and what it meant for our Planet both short and long-term and that is how we will begin our story

Getting back to the cure, if we can find a way to reduce the mid-century forecasted population numbers, how do we do that in a civilized manner and perhaps reduce it even further?

That is the message we pray our leaders will hear, for the last thing we need is some draconian plan of action, for only misery lies down that path. At most, we have only a short time to start taking the necessary steps, for many of the needed actions will take time to implement.

*****

INTRODUCTION

The inputs for this book derived from many of the most brilliant minds on Earth. Unfortunately, they have a habit of using words and drawing conclusions based on an in-depth-knowledge of science beyond anything we learned in school, college, or to the selected few post-graduate readers. The only way to cope with this is to make sure we at least know the rudiments about our Earth before delving into the subject at hand, and please fellow readers, have patience, for not everyone knows everything and there may indeed be some surprises lying in wait for you, and if nothing else, this might be an interesting refresher course, not to mention a check on whether this book is correct!

THE TIMESPAN

One of the more difficult things to grasp are time frames and numbers. As an example, let's take a look at the age of things and where we stand.

Our Universe is stated to be 13,400,000,000* years old. (13.4 billion)

Our sun is stated to be 4,600,000,000 years old. (4.6 billion)

Our Earth and Moon are stated to be 4,580,000,000 years old.

Plant and animal life is stated to have begun 300,000,000 years ago. (300 million)

The dinosaurs are stated to have become extinct 66,000,000 years ago.

The last 'Ice Age' began to end 15,0000 years ago. (15 thousand)

The last 'Ice Age' ended 10,000 years ago.

Farming began some 7,000 years ago. (7 thousand)

*Based on the hypothesis that it is as far back as we can see in light years with the telescopes at hand. I ask you, with even greater telescopes what else might we find? I have no doubt other universes, with the clue being the finding of older galaxies beyond our young and distant ones. If our Universe is expanding, what is it expanding into and how long has that been there, never mind what our Universe came from and the age of that substance? Only then do we perhaps get into the use of trillions!

1. CARBON, OUR PLANET AND LIFE PROTECTOR

My whole career has centered around the carbon cycle stretching from University to Agriculture, Horticulture, the energy business, and now 'Climate Change'. Despite this, my first detailed foray into that subject was inspired by Rachel Carson's 1962 book; 'Silent Spring.' Years later, with so many new discoveries, I decided to take a refresher course on the 'Carbon Cycle' at Iowa State. With its emphasis applied to agriculture, I was still surprised that it omitted most of the part that dealt with how surface rock and its carbon interchanged with magma and the interior of our Planet, something that is critical to understanding how it works. Indeed, when I went on-line and searched for images of the carbon cycle, not one of them included that part and this is what we found.

The Typical Carbon Cycle Charts.

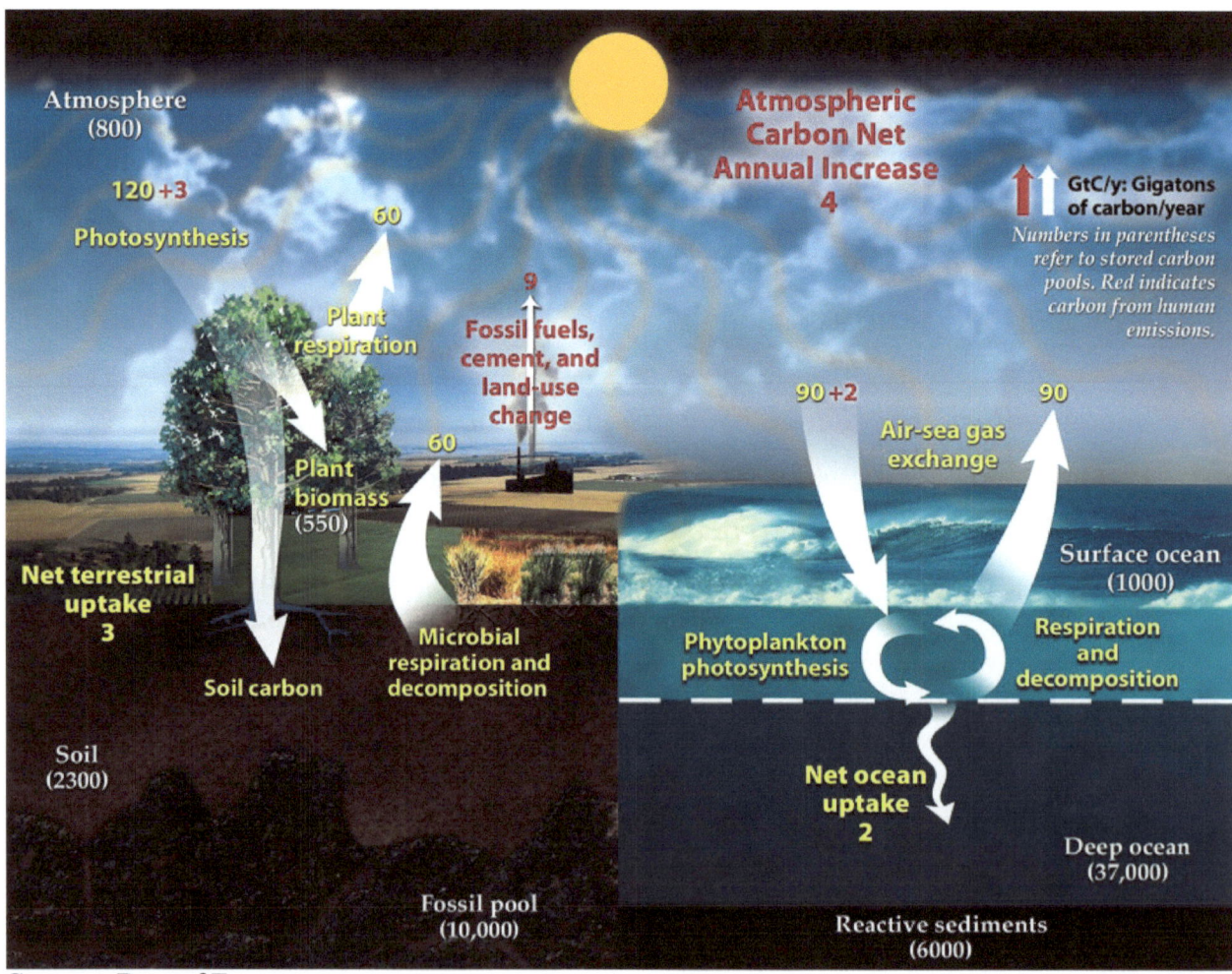

Courtesy Dept of Energy.

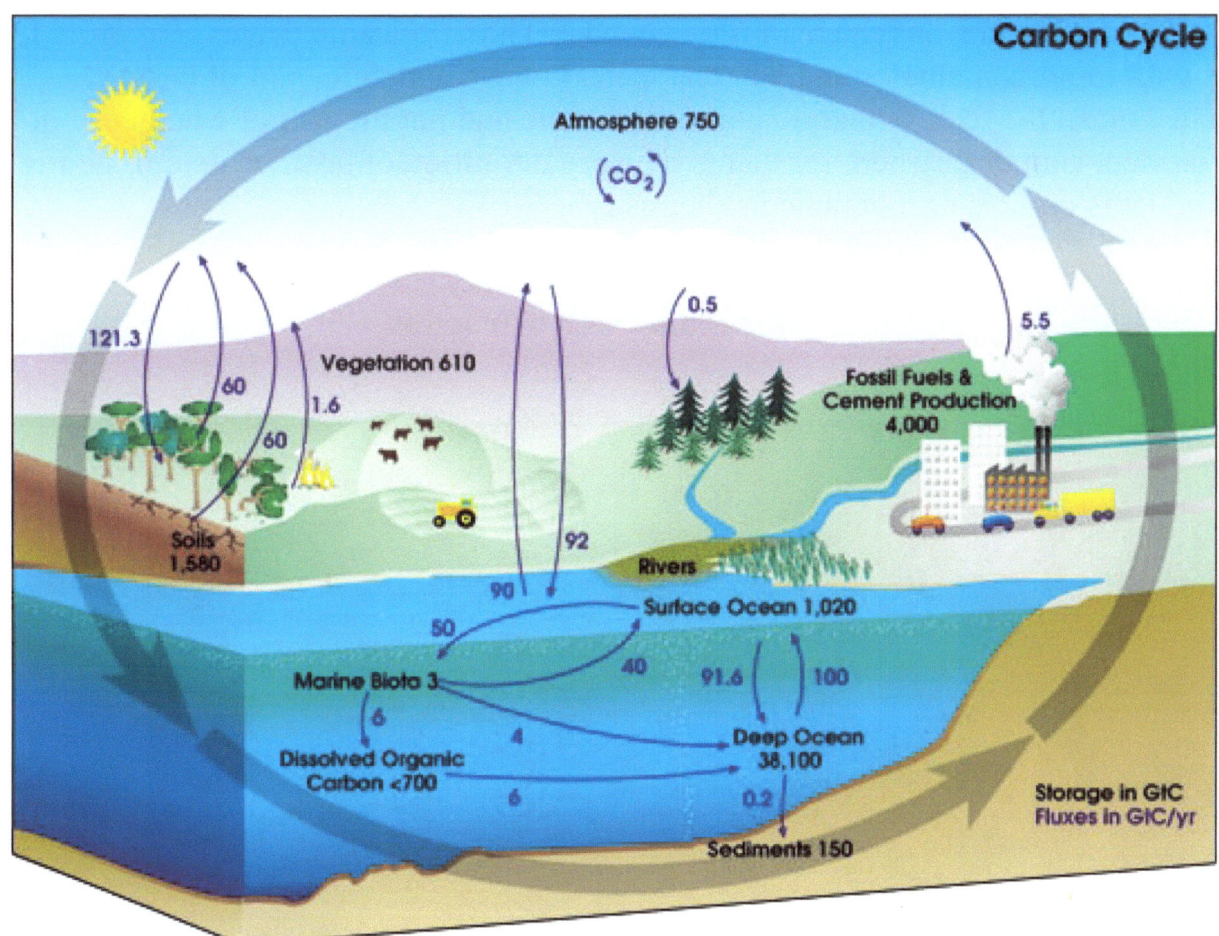

Courtesy NASA. (Both charts show annual and current levels of Carbon in short tons).

And now to the secret; that missing piece is at the very heart of how climate change works and how every Planet has a carbon life-cycle that eventually ends in its death with all living creatures dying from freezing! That death is avoided if intelligent life evolves and in turn manages the 'Carbon Cycle' such that it prevents the formation of 'Ice Ages.' But first, a little about carbon and its place in our World.

Carbon is the fourth most abundant element in the Universe after hydrogen, helium, and oxygen. However, if one had to choose the Universe's miracle element, the hands-down winner would be carbon. In its star and planetary forming stages, it is found in hundreds of forms ranging from the softest to the hardest product (graphite to diamonds). However, given its affinity to bond with oxygen, here on Earth, carbon monoxide and dioxide are it's most common forms. Perhaps it might come as a surprise to some that on our Planet, with the enormous pressures under which our magma or liquid rock is found, gaseous compounds become dissolved in that matrix. Thus, in a volcanic event, when magma is released from these incredible pressures into the atmosphere, the sudden releasing of those dissolved gases creates the massive explosions we associate with volcanic eruptions. However, more importantly, it is the source for our World's supply of carbon dioxide and the building block for the many products that form the organic portion of the 'Carbon Cycle' on which all life depends. Those volcanic events also provide us with important clues about what happens in the 'inorganic' part of the cycle. By that, I mean the rock formation part of the cycle and not the fact of whether or not a product contains carbon, for that is the very definition of how we determine if a product is organic or inorganic in chemistry.

Our Planet's Construction

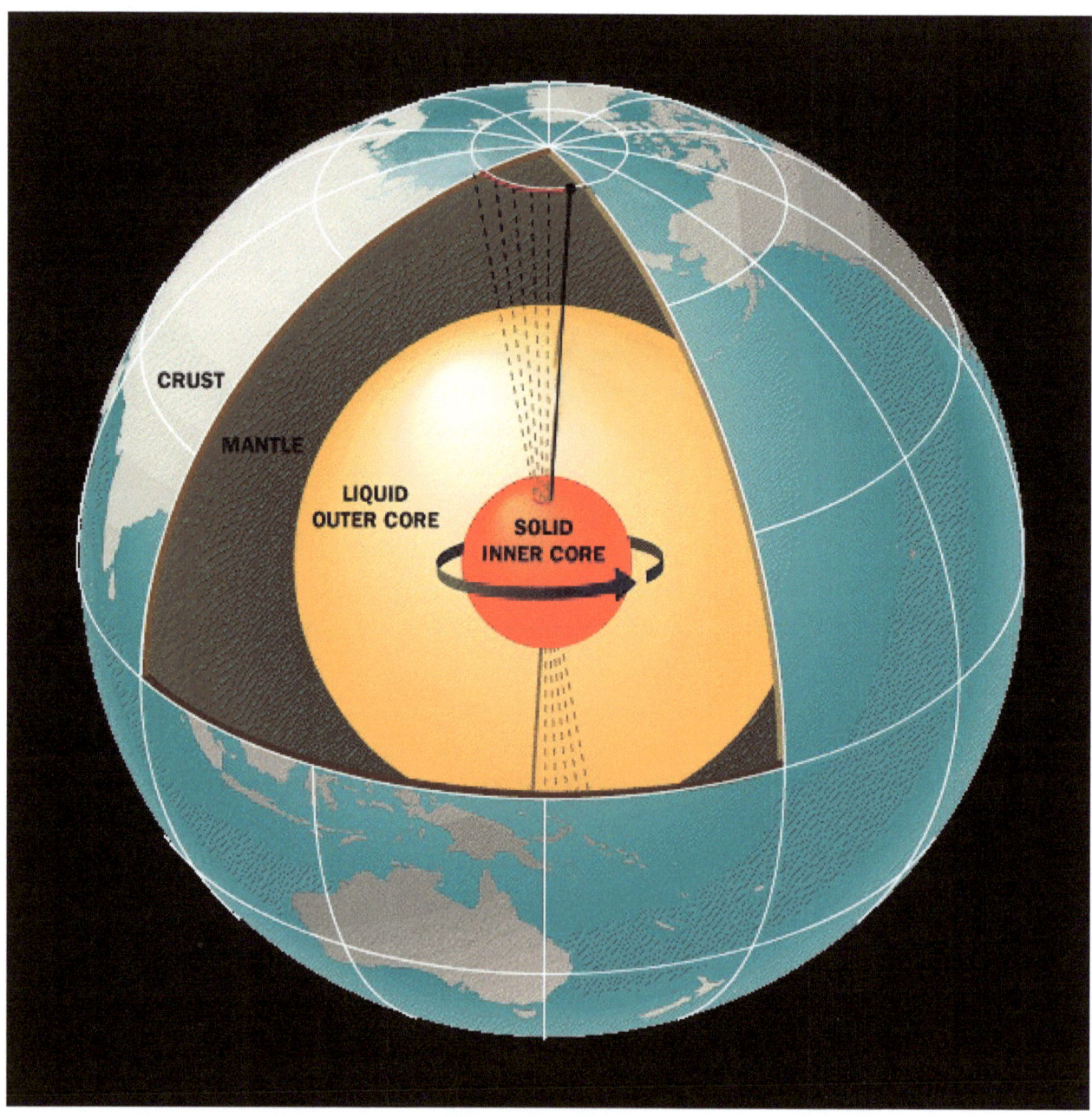

Courtesy RPI, Atmosphere, JPL/NASA, Michigan State 12/1/14 Study, ref Jim Erickson.

But first, there is another clue that volcanoes leave behind. Some of the Earth's biggest volcanic events did not involve explosions, but merely seepage or oozing events for the magma involved came from the Earth's deep core and contained few dissolved gases under pressure. Allow me to explain:

When the Earth was formed, we have always assumed it was from the tail of molten material left behind as our Sun coalesced into a revolving ball of liquid materials. Travelling through frigid space this ball of liquid rock would be giving off gases to form some kind of atmosphere around itself. Scientific thinking is that the primary gases would have been methane (CH_4) and carbon monoxide (CO). The only problem was that with the huge volume of 'Solar Wind' and the absence of any Earth magnetosphere, these vestigial atmospheres would immediately be blown away and be lost into space. With the first relatively hard skin around the Earth to stop this loss not occurring for at least 100 million years, the loss of carbon in these prehistoric atmospheres must have been enormous. Further supporting this theory is the fact that the first clues of Earth's magnetosphere forming do not appear until after this time and so there was nothing to stop our 'Solar Winds' from blowing this atmosphere into deep space..

The Earth's Magnetosphere

Artistic Impression courtesy of NOAA.

With space at absolute zero and our Planet's temperature from core to mantle being 13,000-9,000°F (7,000-5,000°C), the thermal circulation must have been great despite its turbidity. Indeed, the formation of the first crusts must have been a constant battle between gaseous bubbling and how thick the crust had to be before it could stabilize. Thus, unless there was some instrument present that could measure that gaseous loss, this is a job for computers and software writers, and my 90% loss number is probably as good a guess as any.

We now come to the explosive type of volcano whose magma contains large amounts of dissolved gases. So where did they come from? As our Continents move around our globe, so too do the ocean beds drop below our Continents with those beds containing lots of both carbonaceous materials and carbon rocks. As they drop deeper into the Earth's core, so the heat converts those materials back into basalt and CO_2 with that gas in turn being dissolved into the liquid magma around it, and hence the explosivity of that material. Volcanoes near the ocean are therefore usually explosive in nature. The only exception being the ones sitting on top of what scientists call 'Hot Spots' or plumes, which are fed from deep within the Earth and are low in CO_2 and non-explosive in nature.

Our Modern Carbon Dioxide Source

Photo courtesy Ngunyi.

As the Earth ages so too does volcanic activity, if for no other reason than the Earth is cooling and it's surface crust thickening. This means that our source of CO2 also is declining with its up and down cycles depending on volcanic activity. Scientists currently estimate the carbon on the Earth's surface to be 6.5 % of Earth's total.

So where is our CO2 going? First it dissolves in water, both fresh and saltwater to form carbonic acid, a weak acid to be sure, but one that reacts readily with the calcium shells of plankton and other sea creatures forming limestone and marble type rocks. What little is left in our atmosphere provides living creatures with the many forms of carbon necessary to support life in it's many forms, both animal and vegetable.

The Fossil Fuel Producing Periods

Chart Courtesy of Wikipedia.

And this everyone (Including our scientists!) is the natural cycle of Carbon and how all earthlike planets will end up with none on their surface and therefore becoming devoid of life.

Earth, a Planet in Peril

The proof of all this can be found in Earth's 5 previous 'Extinction Events' with all of them beginning with high levels of CO_2 from various reasons and ending up with almost none due to being recaptured by reaction with alkaline substances to form long term rock, and in a bizarre twist of fate, massive amounts of vegetation which was the basis of nearly all of our fossil-fuel.

In theory, this is how all surplus carbon gases end up with all planetary life ending in this way, even assuming they were in a life supportive orbit around their sun and had a magnetosphere in the first place. Put another way, life on Earth has only existed for 0.000002% of the life of our Universe with little or none of that evidence from elsewhere able to reach us as yet! (Someone please check that math for me!)

To answer that would have been impossible even twenty years ago with the research done since then almost beyond belief. However, we need to know the key conclusions about this research for it, in turn, helps us to complete that part of the carbon cycle. Based on this work being conducted on the oldest rock fragments found on our Planet, we can date the key events in the formation of our solar system as follows: The carbon held in our planet's crust, mantle and core not only dominates the carbon cycle, it also dictates our long-term carbon dioxide levels and in turn World temperatures, sea levels, and via the unique positioning of our continents around the North Pole, the oscillation of our warm and cold periods dominating to-day.

Adding our own carbon cycle numbers (the crust and mantle) to the above, we end up with the following adjusted cycle amounts:

Our Planet's Carbon Cycle

(All data in metric tons of elemental carbon.)

The Planet's original core, mantle, and crust.	4,360,000,000,000,000,000
A 90% loss of carbon into space during the time before a crust and magnetosphere formed–my guess.	(3,924,000,000,000,000,000)
Balance	436,000,000,000,000,000
Earth's crust (6.5%)	(283,000,000,000,000,000)
Earth's mantle (Δ.)	152,99999999999934595
Atmosphere	810
Dissolved in water	36,000
Biosphere	1,900
Coal in the ground	900
Low grade coal	18,000

Oil	150
Natural Gas	105
Other fuels (Shale etc.)	540
Methane	3,000
Limestone, marble, etc.	4,000
Of which fossil fuel	26,695
Of which the biosphere	38,710

(Please note, still using Govt. data.)

This table shows that the Carbon in the cycle, on which our lives and civilization depend, is only a minute portion of the amount in our Planet. It also indicates that humanity is in the process of moving fossil fuels, from over 300 million years in the past, and which took 70 million years to produce, into modern times. Even more stunning is that we may be consuming it in less than one thousand years. Little wonder that the atmospheric levels of CO_2 on our Planet are out of balance. The question is what is the best amount to support life on Earth?

For that answer, we have to take a look at what we might learn from history.

Carbon is a naturally occurring element with a radioisotope reading of 12 and known as C_{12}. On exposure to sunlight it converts to C_{14}. Over time, and in the absence of sunlight, it returns to its original form of C_{12}. (in 60,000 years). With a half-life of 5,730 years, this means that we can tell if that carbon came from a surface source or from deep in the Earth where sunlight is absent. Thus, we can tell where the carbon dioxide in our atmosphere originated, a vital factor in identifying the source, for almost all fossil fuels were laid down millions of years ago and are found only at great depth on our Planet's surface and so consists of carbon in its C_{12} form.

The petroleum industry defines the carboniferous period as including the Pennsylvanian and Mississippian periods covering a total of nearly 70 million years and some 300 million years in the past.

This was a classic example of the short-term CO_2 cycle and how it ends. Conditions in Triassic Earth (Pennsylvanian, Mississippian and Devonian periods), were ideal for the luxurious growth of plants: a moist, and warm atmosphere, rich in carbon dioxide, so rich in fact that the plants literally pulled all the CO_2 out of the atmosphere leading to a period of cooler, dryer and more arid conditions followed by an 'Ice Age'. While most people are under the impression that all fossil fuel was created and laid down in the Carboniferous era, that is not true, with coal and natural gas being formed after that time. Thus, the search for fossil fuels is not quite that simple.

Carbon dioxide is produced when organic compounds such as fossil fuels are burned. There are three phases to that burning process; the ignition or warming phase, the burning of the plant material to carbon, and lastly, the burning of the carbon itself when it combines with oxygen to form carbon dioxide and is the 'high heat' phase of the burning process.

With Volcanic activity the main source of CO_2, before modern civilization, its current source from burning fossil fuels is far different with that source mostly CO_2 and not the typical mix of CO_2, SO_2 and NO_2 from volcanoes that in many ways had off-setting influences on climate. That toxic brew of Carbonic, Sulfuric and Nitric acids reacted quickly with rocks to become neutralized whereas today carbonic acid neutralization is a slower and longer process, and with the absence of the cooling gases, having a greater impact than normal on upward temperature movement.

To-day's situation requires a different answer with the good news being we can do it with it playing a bigger role than our experts have forecast!

Returning Carbon Dioxide to the Soil (Sequestration)

Farmers have long understood that new soils are rich in organic matter and produced the biggest crops. In N. America yields of sixty to eighty bushels of wheat were common when the first crops were grown, yet within twenty years had dropped down the thirty to forty bushels per acre level. Some though, and there was only a handful in the whole of Western Canada in 1958, knew the value of a mixed rotation coupled with livestock farming. Field days on these farms showed they were still raising crops yielding the same sixty to eighty bushels they or their fathers had achieved some forty to fifty years before when their land had first been broken. Soil tests showed the same high organic content from those days (6-7%), and with higher nutrient and moisture content explained their success for all to see. The farmers, without the grass legume (brome alfalfa) rotation, were doomed to lose organic matter from their soils. Usually, because they were unwilling to raise animals (too much trouble) and lose three years of grain production revenue. (While the grass legume replaced that organic matter, for without animals they had only a limited market for hay), and in the end, added back ten to twenty bushels of yield by using fertilizers, weed sprays, better varieties, and fallow years to replace depleted moisture. With the advent of 'no till' farming, the annual losses of organic matter were stopped or slowed to only a trickle, with everyone joyful at that accomplishment. There was only one problem; it did not solve the problem of rebuilding the soil to a more productive level.

In all of this nobody had raised the issue of where the lost organic matter had gone. Into the air as carbon dioxide! With the alarm sounded about 'Global Warming,' the first question raised was what had been the source of that carbon dioxide? By tracking the different isotopes of carbon ($C14$ from our planet's surface and $C12$ from fossil fuels), it turned out that 20% was coming from farming and

80% from fossil fuels. However, on a cumulative basis, with agricultural emissions stretching back thousands of years, the split was fifty-fifty. The next question was, could farming somehow recapture back some of that lost carbon (sequestration)?

Well, we already know that answer, add back a grass-legume crop into farming's crop rotation system, adding three years to a two-year cycle making it now a five year cycle. The short term consequences would be negative, but longer term, with higher yields, partially offsetting some of that lost production. Texas A&M studies show that after just three of these rotations, soil organic matter would have been raised from three to six percent. The key being to keep the legume active and the rotation being three years in length. Everything considered, a remarkable achievement if done. The calculation by the EPA was based on what would happen in the top six inches of soil. Unfortunately, it should have also included the thirty inches below that where the grass and legume roots would reach and be returning some of that carbon dioxide

Our scientists have been stating that our soils are being worn down and have released about half (3%) of the organic matter they originally contained into the atmosphere as carbon dioxide. That number is wrong and here is why. When land is first broken for farmland, and especially true for the Western Plains of the USA and the Prairies of Canada, plant roots typically are very dense for the first six to nine inches of soil, and then after that quickly thin out until six or more feet is reached when no traces are found. Soil tests typically are taken for the top six inches of soil as it contains most of the nutrients the plant will need with only 1% or less any deeper. And this is where the problem lies when measuring soil carbon content.

As a general rule for fresh land, we used a number of 7% total dried organic matter with closer to 9% for heavier soils in river-bottom lands. Other Countries and their numbers were in the U.K. 4-12%, Denmark 10%, Belgium 15%, Iran 1-10%, the USA 6-9%, Columbia 9%, Brazil 2-12%, Malaysia 2-15%, Indonesia 2-10%. Now, as a soils expert, let me blow the lid off our non-soil scientists, for this is only the tip of the iceberg in understanding our soils.

I have to admit that I count myself as one of the guilty ones in measuring the organic content of our soils, for to start with, just about all soil sample cores only go down 6-9 inches maximum into the soil, with typically only the first six inches being bagged for actual testing and measurement. Only when digging trenches (for farmer field days and typically of 3 to 6 feet in depth), does one get a true picture of the profile of a soil. The reality of original soils is a profile of dense organic matter for the first three inches, then about half that for the next 3, then a rapid drop off to 1% or less if deep plant roots can be seen. Again, there is a trap here, for guess what, the carbon (humus) at these depths is often higher than what one would guess, for it builds up over the centuries and yet can be invisible to the eye.

What happens is this. Over the hundreds of years of previous growth, plant roots have broken down into the carbon skeletons of their cells. These carbon skeletons are extremely light and fragile and are impossible to see with the naked eye. And here, let me explain what the problem is.

If one were to take a soil sample in western Canada, it gets confusing, for glaciation scraped off all the ancient soils and replaced them with lakebed deposits, and other stirred up mixtures, and are less than ten-thousand years old. Regardless, most of the Prairies have readings of 5-7% for the top 6 inches of soil, then 1% for the next 30 inches. Thus, instead of reporting the organic content of the soils as being 6% (for 6 inches), it should have been reported as an adjusted reading of 11% equivalency if including the organic material contained in the 30 inches of soil beneath it. I know it sounds crazy, but how else should one measure the amount of organic matter held in the soil, for it should be the total that counts? It was a man by the name of Bill Lobay, in Edmonton, Alberta, who ran the Soils Lab there, that I credit with bringing the true understanding of how soils behaved and worked.

It was while running some experimental soil tests for me, that when he was conducting the percent by dry weight measurement, there would be intermittent bright flashes, especially towards the end of the soil sample's time in the electric oven, when the heat was rising to the six-hundred-degree centigrade range. Bill's explanation was that by that time the organic matter had burnt off except for the elemental carbon in the soil, which only ignited and burnt to ash at above the four-hundred-degree C. level.

"You mean the humus'? I asked without thinking.

"Why yes," he replied. Only later did I remember that I had failed to pay attention to the twinkle in his eyes, indicating that he knew something else of importance, which luckily, I had the chance to confirm later.

"You mean when we are measuring organic matter we include elemental carbon which is not, in fact, organic," I asked?

"I suppose that is correct, but the logic of doing that, is the source was from organic material, and burning off all carbon the only practical way to express that content in soil," he replied.

A day later, and back at the Vegreville Experimental Station with the soil sample results for Dr. Ross Cairns, its Director. Relating to him the story of the humus in the samples, he proceeded to give me a lecture on humus that in turn, he credited to the Ukrainian farmers in his area that knew all about it from their homeland. Still vividly remembered, for somehow it was the thing that allowed me to complete my education into soils and the role of humus and fertility in that equation. What he taught me that day was how the carbon in the soil was nature's amendment that more than doubled its capacity to retain and hold both nutrients and water, leading in turn to an almost doubling of yield in his test plots at the Station. Indeed, he was hopeful that it might lead him to his problem of returning good tilth to the solonetzic soils in that area. (Remember the slithery and gooey soils that bogged down the advancing Panzers when they invaded Russia in WW II, well, those were the solonetzic soils of the Ukraine.) Thus, carbon plays two roles in soils. As part of the molecular structure for the organic matter in our soils, and as microscopic particles of carbon that also cling onto water and plant nutrient materials.

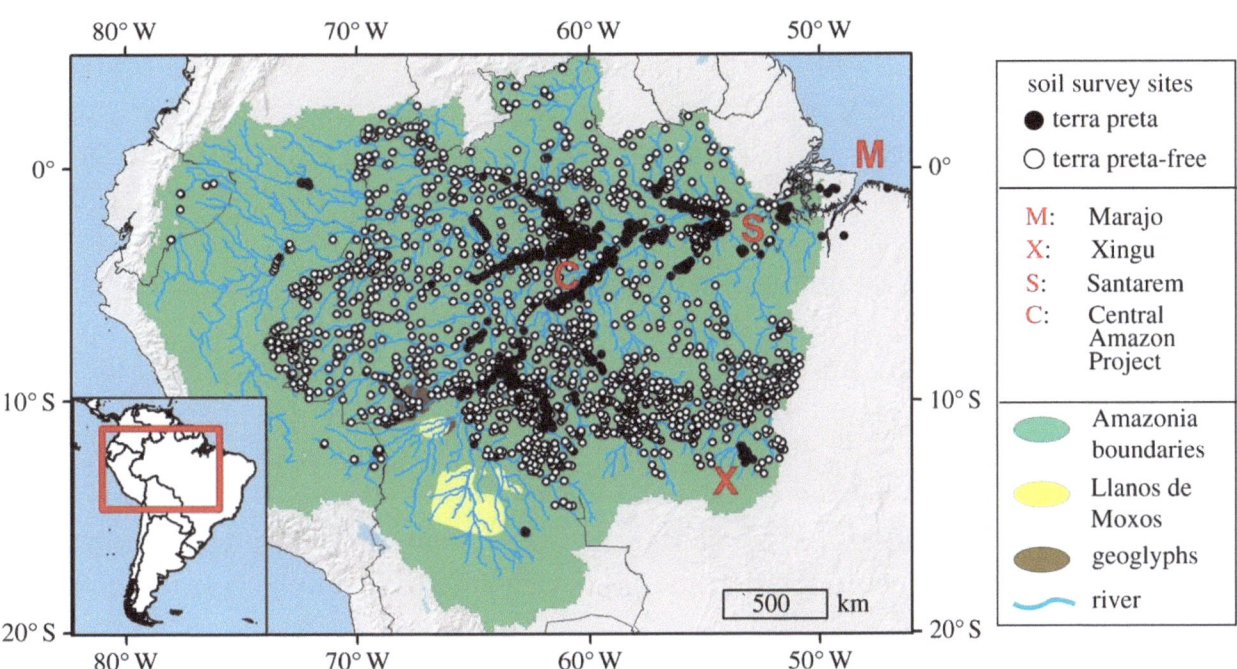

Image Courtesy of 'Proceedings of the Royal Society, C H Mc Michael et al. 1/8/14.

This second category is particularly important when discussing forests and their natural cycles when they involve burning, but the area that is absolutely critical is when we talk about tropical or lateritic soils and their management. Under these conditions, where tropical rains wash out anything soluble from soils, carbon or charcoal becomes all important for it holds on to soil nutrients and water to tide plants over the typically long periods without rain. Its role is not as a nutrient, but as a soil amendment, for when added to soils, dramatically improves yields. Thus, in the tropics, when clearing land by burning the trees and driving off carbon as CO_2 into the atmosphere, that practice is not only wasteful, but ruins rather than improves topical soils and should be regarded as a destructive action. When one thinks of what has been done in the Amazon and in Indonesia, the question is why? These were people with only two thoughts: making money and doing it as quickly as possible. The tragedy in the Amazon was compounded by the fact that the ancient settlers there knew about the value of carbon and actually made some of those carbon rich soils which they call 'Terra Preta'.

In plain English, what the above lesson teaches us is that the 1% organic matter at depth is mostly carbon and not organic matter, yet in the breaking down process, while in the top six inches, the reverse is true. And that is why, when calculating the carbon held in soils, it is crucial to know the difference between the carbon content and the other vegetation, something which is not measured in soil samples, and the key to understanding how much carbon can be returned to soils.

In the late 1970's, zero till was touted as the magic way of minimizing erosion while at the same time reducing tillage costs. Well, it does, but again researchers kept looking at soil organic matter as if it were all of one type and not, in fact, changing in front of their eyes. Indeed, if humus and roots go together, it does little good if it is all in the first 3 inches of the soil. Soils with zero till being practiced typically show an organic matter content of only 3.5% and no way to build it up. In other words, surface till is great, but plowing after a legume grass crop is a must with a return to no till after that also a must. For non-farmers, the reason for this is that humus is the left-over carbon skeletal frame of a plant after it has rotted away. These skeletons are extremely delicate and light and can easily be washed or blown away when exposed to the air. Hence when lost or added to soil, it changes its quality. As for nitrogen loss, have our experts never learned the basics, for if broadcasting; use ammonium nitrate, for the plant roots absorb it in that form, whereas urea has to be broken down by bacterial action into ammonium nitrate before it too can be absorbed, thus losing ammonia into the air. Likewise, if doing application on broad-leafed plants, use urea ammonium nitrate solution for that is the only form of nitrogen that the plant can absorb through its leaves.

What does all this mean? The Earth's soils, before man's appearance, contained, (all data expressed as the total dry organic matter as if it were contained in the top 6 inches of soil) 11% by weight with 8% of it as humus or carbon and 3% as true organic material in the process of being broken down into carbon. This means that if we wished to return our soils to their original levels of carbon, we would have to add back 8% by weight expressed as carbon equivalent. But, and I cannot stress this strongly enough, that is not the optimum amount of carbon that the soil contains which is another 8% higher if, in the form of charcoal, and not humus, which is too easily lost due to its light weight, even if using best farming practices such as 'no till'.

Using World Bank/FAO arable acreage numbers and EPA's carbon conversion/acre formula (21.21), this equates to a carbon sequestration potential target for N. America of 146 billion tons of carbon equivalent, and for the World, 886 billion tons. Assuming a 50% success rate, this would be sufficient to solve our problem, but not if we continue to spew out the 40 billion tons of CO_2 from energy as at present.

N. American Arable Acreage Carbon Recapture Potential

(CO2 expressed as billion tons of carbon equivalent)

(Assuming a North American Alliance of Canada, Mexico and the USA)

(FAO, EPA data, 0.6 billion arable acres x 21.21 tons of carbon per acre x %)

Years	% Organic Matter	% Humus	Cum. Total*
5	1= 12.726	0.5= 6,363	19,089
10	2= 25,452	1.0= 12,726	38,178
15	3= 38,178	5= 19,089	57,267
20	4= 50,904	2.0= 25,452	76,356
25	5= 63630	2.5= 38,178	101,808
30		3.0= 44,541	108,171
35		3.5= 50,904	114,535
40		4.0= 57,267	120,897
45		4.5= 63,630	127,260
50		5.0= 69,993	133, 623
55		5.5= 76,356	139,986
60		6.0= 82,719	146,349

Rounded up to include pastureland under rotation. Assumes all charcoal applied by slurry injection at 6 to 9-inch depth.

*Sequestration cumulative gain in each five-year crop cycle period.

If the rest of the world followed the lead of North America, the numbers would be:

World Arable Acreage Carbon Recapture Potential.

(CO2 expressed as billion tons of Carbon equivalent.)

(3.8 billion arable acres x 21.21 tons of carbon per acre x % =)

Year	% Organic matter	% Humus	Cum. Total*
5	1= 80,599	0.5= 40,299	120,897
10	2= 161,196	1.0= 80,598	241,794
15	3= 241,794	1.5= 120,897	362,691
20	4= 322,236	2.0= 161,196	483,432
25	5= 402,795	2.5= 201,495	604,290
30		3.0= 241,794	644,589
35		3.5= 282,093	684,888
40		4.0= 322,392	725,187
45		4.5= 362,691	765.486
50		5.0= 402,990	805,785
55		5.5= 443,289	846,084
60		6.0= 483,588	886,383

*Sequestration cumulative gain in each 5year crop cycle period.

Please note that in the above tables, the 1.5% increase targeted for the first five, 5-year periods, represents a large amount of material which will require occasional deep cultivation along with current zero till practice. As a general rule, organic matter breaks down into humus in a 4 to 7-year period (tropical-v-normal conditions), and hence is achieved in one annual application of harvest trash or grass/legume plow-down.

The FAO estimated the world's Forests to be holding an additional 638,000 billion metric tons (tonnes) of carbon.

The Timing.

While on this subject, now is as good time as any to explain something else about farming to not just American city dwellers, but to all the people of the world. As industries mature and become more efficient, so too does the cost of improving that further. Perhaps not obvious at first, the lowest cost improvements are always made first, with the most expensive last, and farming is no exception to that rule.

Classified under our antiquated tax laws as a family business, farming families were subjected to standard death duties such that roughly anything of value over two million dollars was taxable. Thus, few farms could be passed on to family members and have instead been sold to large investors. As a consequence, any efficiency improvements that take large amounts of capital have been deferred and, in most cases, were discarded as being impractical. Buried in our revised tax laws of December 2017, is a provision for a tenfold increase in that tax deduction, which overnight has suddenly opened the door for the return of farming families to carry on the tradition of passing on these businesses to their immediate family and to once again make expensive improvements practical. If ever there was a time to make family farming once again an attractive business for humanity, the timing for this is perfect with a new farming revolution about to dawn in America, and in turn a revitalization of rural life and their communities. If anyone in our Government reads this, what an incredible opportunity it would be to write something into a program to encourage our farmers to sequester carbon, not to mention renewable energy.

Is anyone in government listening? Eight months later the silence is still deafening!

*****

2. OUR POPULATION

Homo erectus is stated to have appeared here some 250,000 years ago.

Homo sapiens is stated to have appeared here some 130,000 years ago.

The last true super volcanic eruption occurred 72,000 years ago reducing homo sapiens to less than 200 family units. (With some debate continuing whether it was from that eruption or something else.)

The Neanderthal species died out or merged with Homo sapiens some 33,000 years ago.

The industrial revolution began 200 years ago with our population reaching 1 billion.

Our population is now 7 billion. By 2050 our population is expected to be 9.7 billion[1].

Our life expectancy is currently 78.6 in the USA and 83 plus in Japan.

*****

The above has been largely driven by the fact that in our poor and underdeveloped countries the birth rate varies from 3 to 6 children per family unit. However, as those conditions improve, that rate has dropped to the 2 per unit level thus lowering the growth rate substantially.

Mathematicians, particularly over the past one-hundred-years, have pointed out the phenomenon of how growth rises in proportion to the percentage of child bearing young in a population, and in turn, allows them to predict the kind of exponential growth that ensues. Despite this, many in our social media have been pointing out with glee how, with better education and an improved standard of living, the World's population problem has gone away. While their audience might leave feeling better about such a depressing subject, the disservice done to the World and its future is a crime of the first magnitude and they should be bowing their heads in shame for using such a cheap tactic, all in the name of the ratings and popularity game.

The fact is that no matter what our politicians and leaders say, our World population is still climbing and is expected to do so through this entire century.

This alarm is different, for it is based on the physical damage humanity is doing to our Planet, and to what is waiting for us around that corner.

---

[1] The United Nations delegated the task of keeping track of the world's population to the World Bank.

The 2018 World Bank Population Forecast

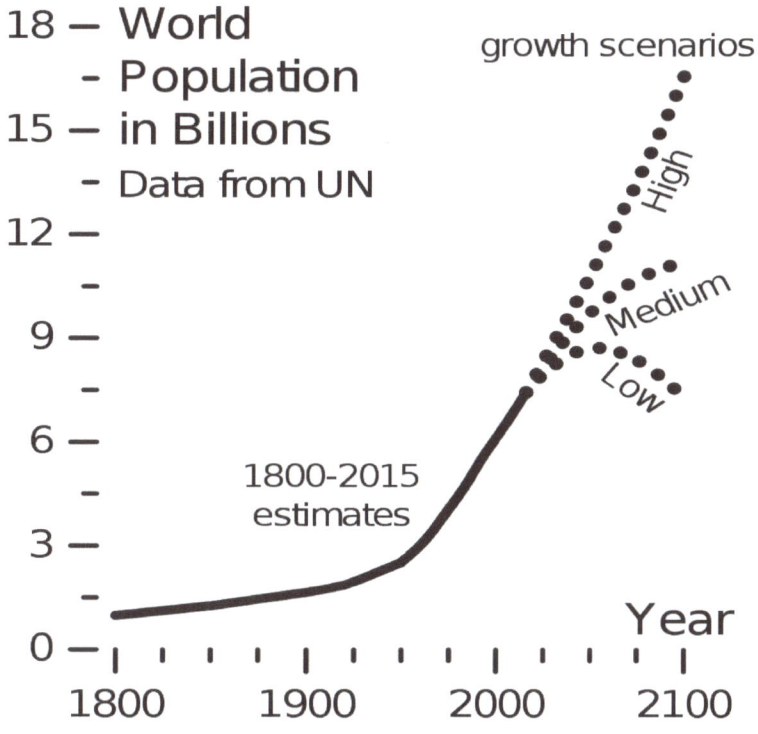

The Split by Continent.

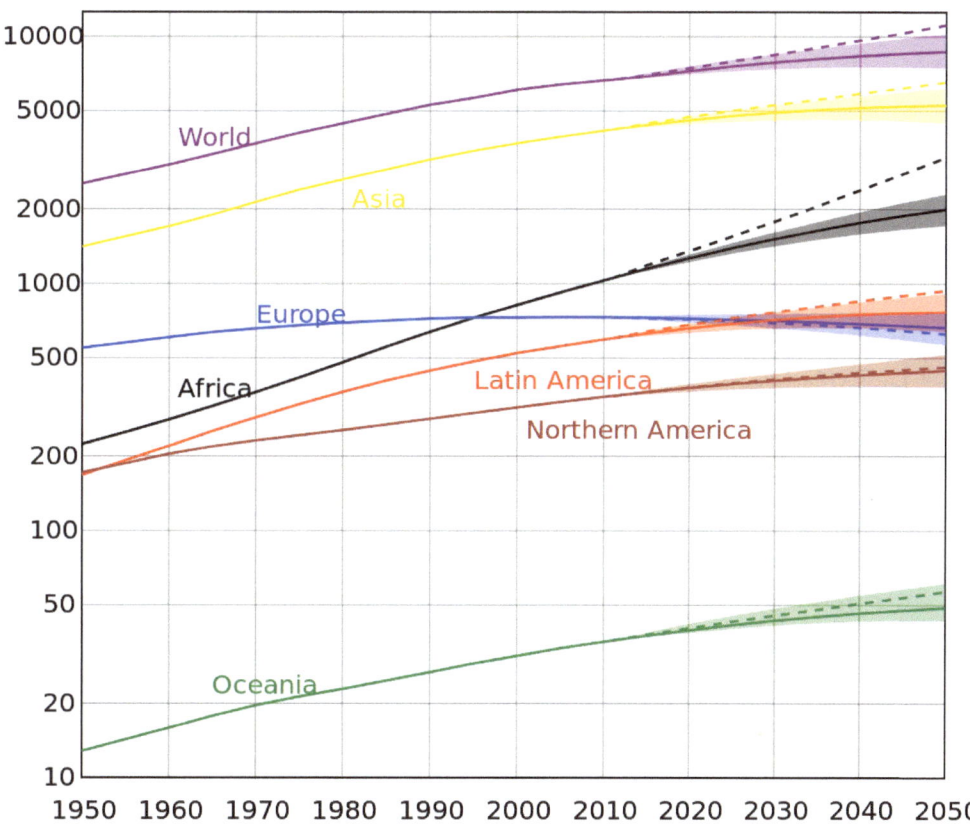

Charts courtesy of World Bank.

Earth, a Planet in Peril

To understand this and what has been happening, its consequences and the gradual build-up that has been occurring, a short trip through history is helpful.

Early man faced an extinction event almost before they could get started. A truly giant volcanic eruption occurred in 74,000 BP (Before Present), that according to geneticists reduced our worldwide population to less than 200 family units. It was a super-volcano known as Mount Toba on Sumatra in the Indonesian archipelago with ice-core records in Greenland and Antarctica, backed up by the local geological evidence, indicating its size. It was so large, and leaving a sulfur dioxide (SO_2) signature of several thousand years, indicating its plume may have reached into outer space. As for the cooling signature, for SO_2 does that, it dropped world temperatures to their lowest temperatures during that Ice Age, with its marker showing up in the ice-core charts.

The traditional way of putting this into perspective would be to compare it to the eruption of a volcano in 1980 in Oregon, known as Mount St. Helens, which had an ejecta volume of 0.25 cubic kilometers (km3). Toba's was 2,800 km3, making it 11,200 times greater in size and a true super-volcano. Even the Yellowstone eruption of 600,000 years ago, was smaller, so we cannot dismiss this eruption as being part of the World's ancient past and something that could not happen today, for both these volcanoes are still active. Finding that comparison hard to picture, how about imagining if it happened today? Guess what, 'Global Warming' would no longer be a problem with the worry being 'Global Cooling' and is anyone even alive?

That story now brings us to why Africa was our homeland for so long. The eruption of Mount Toba, and its killing of most life, postponed the date of population pressure and the delay of migration until approximately 40.000 years ago. The fact the 'Ice Age' was still active also explains why those emigrants kept moving eastwards until they were once again able to turn South and into warmer weather.

The Ice Age phenomenon, which mainly affects our northern Hemisphere, did not ease until 10-15,000 years ago, at which time emigrants settled in the Middle East area. Meanwhile, by 1880, the World's population had grown to one billion people. However, with the age of discovery and invention, that same population was getting ready to become the fastest growing species on Earth. By 2017 it had reached seven billion, and with a young population of child-bearing people, was getting ready to expand to almost ten billion by 2050, and eleven billion by the year 2100, and this despite the world's birthrate dropping to two children per couple. While alarm signals are being sounded about such things as carbon dioxide atmospheric levels and rising temperatures, another factor was becoming increasingly apparent, damage to all other forms of life indicating that not only were population levels increasing to more than the Planet could support, they were already well past that level, with the Planet's environment already in a steep decline. Indeed, everything indicated it was heading into an extinction event worse by far than that of the Mount Toba eruption some 72,000 years ago with the current loss of plant and animal species at the highest levels ever recorded even compared to our Planet's five previous extinction events.

In plain English, this is what our scientists are telling us. By the year 1900, the planet was already undergoing massive physical damage to its environment with the finger pointing at population levels of one billion perhaps being too much!

By the year 2000, atmospheric signals were predicting the Planet was heading for trouble from what a population of six billion was causing.

And here we are discussing what might happen with ten billion on our Planet in the year 2050!?

The history behind birth control is a story unto itself. All life on Earth has built into it a drive to procreate and multiply to preserve its species. Indeed, even non-cognizant life does so by the system of laying millions of eggs at a time. Only in humans do we find a conflict arising from the fact some are already understanding that overpopulation is now a danger. Thus, Europe as a whole has reached a stable population with zero growth forecast over this period. Even the World's religions have recognized this fact with their dictums neutral or ambiguous on this issue with one exception. The Roman Catholic Church still mandates that no birth control or abortion devices shall be used. I used the Church's old name deliberately, for it was the West's first Christian Religion and was founded in the time of Rome's first Emperors. In its early days, it recognized the role women played in the family and the influence they wielded on their thinking. In the environment surrounding the church two other things were also important to understand. The first was the attitude of the Roman Emperors. They were against Christianity and put to death many of their adherents. Thus, the Catholic Church encouraged large families as the way to gain desperately needed followers. Written into their dictates were rules against abortion and the use of preventative birth devices. With celibacy required of its Priests, the church at the same time recognized the common practice of homosexuality among Romans and turned a blind eye to this practice. Under Emperor Constantine, the male dominated Priesthood banned and downplayed the role of women while at the same time endorsing the banning of abortion and the use of birth control devices. While Jesus had encouraged his followers to 'multiply and go forth,' he never stipulated the above bans. Thus, the Catholic Church's rules are policy rather than religious dictates. The question is; will that Church change its dictates in the face of overpopulation, 'Climate Change', the destruction of our environment, and even extinction? And we call ourselves 'Homo Sapiens'?

By all means we should rely on education and an improving standard of living to set population growth rather than some other uncivilized method of control, but if we are already beyond the level our Planet can support, what should we do?

Nothing but a 'Manhattan Type' program will suffice. By that I am suggesting that the entire World should gear themselves up to the launching of crash programs to educate and improve the standards of living as fast as possible in the countries that need immediate help. That program will have to make sure that all hostilities have stopped in those areas with the factions either in compliance or neutralized. Once done, then missions to educate and improve its citizenry can then be implemented on an 'over-night basis, for nothing else will suffice if we are to gain control over the World's population growth rate.

Make no mistake, World population is the problem with all else merely symptoms. We have three decisions ahead.

–Dramatically cut back our population.

–Reduce our footprint on our planet.

–Some combination of the above.

How do we cut back our footprint? Recycling, switching to renewable energy, efficiency improvements, elimination of waste, conservation and protection of other species. But most of all, immediately raising the standard of living in the developing world.

*****

3. RESEARCHING CLIMATE CHANGE

Our Planet Earth

Courtesy NASA.

Despite 'Climate Change' being a symptom rather than a misdirected problem, the value of learning how our World's climate has been reacting, has been an essential learning process and needs to be told here.

Modern man and our history can at best only go back ten-thousand years (10,000), for before that it was an 'Ice Age'. In other words, our civilization demands we live in a 'Warm Period' and not a freezing Planet, for although only a third of our Planet was covered in ice or snow, another third was too cold to support life as we know it, something that many people seem to forget or ignore.

Earth, a Planet in Peril

With modern humanity having such a short time on Earth, ours has to be a tale of investigating the land and sea around us and what it can teach everyone about climate through the ages. Bringing to bear modern technology, it is safe to say we are learning the ins and outs of climate and what is driving it in modern times. Not so with history, for without anybody around to record what had been happening before three thousand BC, we will have to rely on what our scientists can tell us from the clues left behind.

An exciting story in itself, that tale would make any detective proud. I think I am safe in saying that the first prize should go to our Glaciologists, who in Greenland drilled down in the deepest part of the ice sheet uncovering the annual history of snowfall, which in turn had left a record of temperature, amount of snowfall, the dust and gases it contained, volcanic activity, and many other pieces of information. But most importantly, that record went back 400,000 years (GISP2 ice core data). It was now 1993.

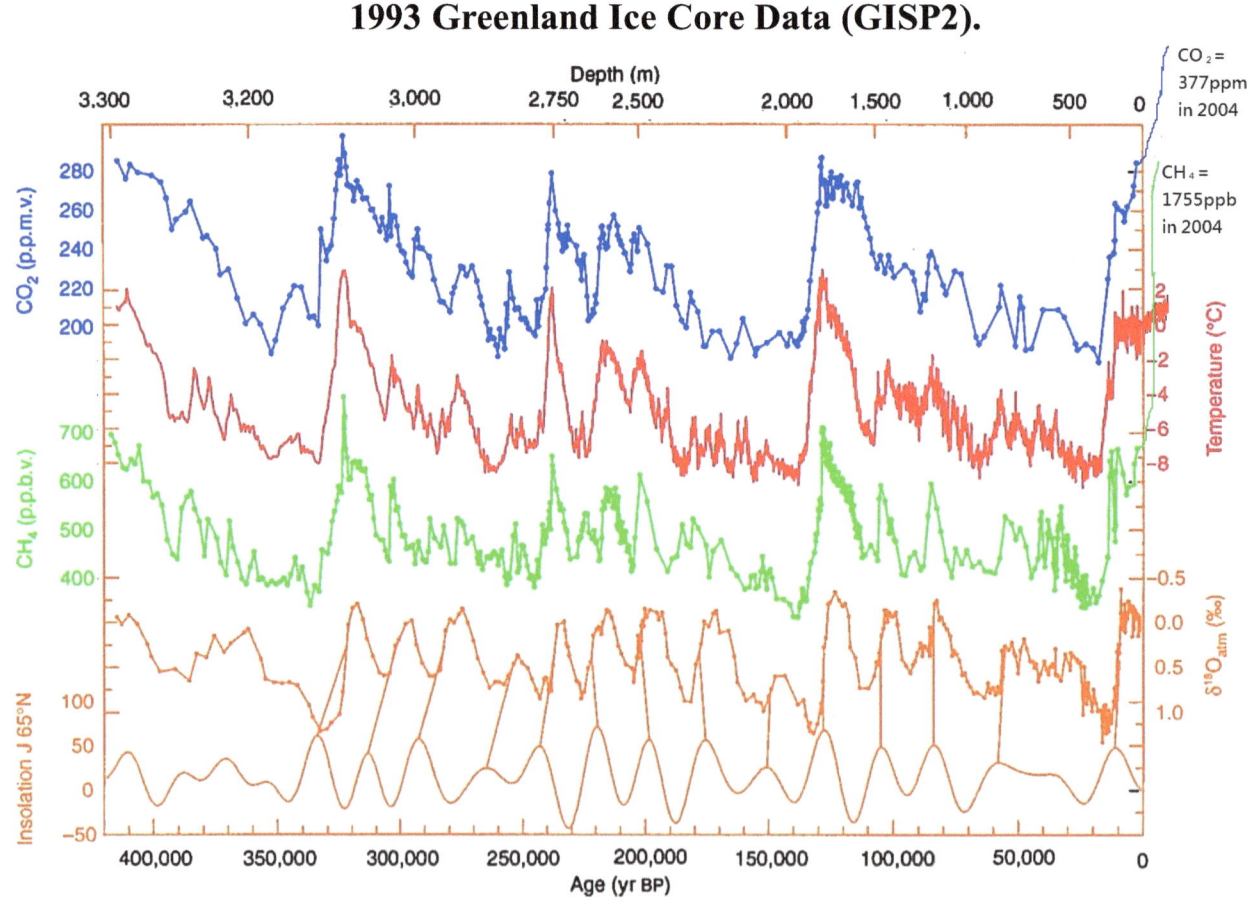

Chart Courtesy of NOAA.

Our scientists could not resist adding the latest carbon dioxide and temperature readings to this chart, presumably to frighten us into believing calamity lay ahead. To me, the big question was what about all those 'Ice Ages'? Not done, the

Glaciologists chose Antarctica for their next drilling site. Choosing an inland area where the ice cap was deep, the site was called Vostok, and also covered 400,000 years of history and confirmed the Greenland data. There was one surprise in all of this, a huge Lake lay under this ice sheet. It was at this point that I became confused and somewhat skeptical of the message being given about 'Global Warming' and Al Gore's presentation of "An Inconvenient Truth". The gist was that carbon dioxide in our atmosphere was causing temperatures to rise with the link between the two proven and confirmed by the ice core data from the above programs. But little was said about 'Ice Ages' nor how our warm periods seemed to trigger a sharp drop into long cold periods. Indeed, that was not explained and there was the gap in our knowledge. It was now 1998.

1998 Vostok Ice Core Data.

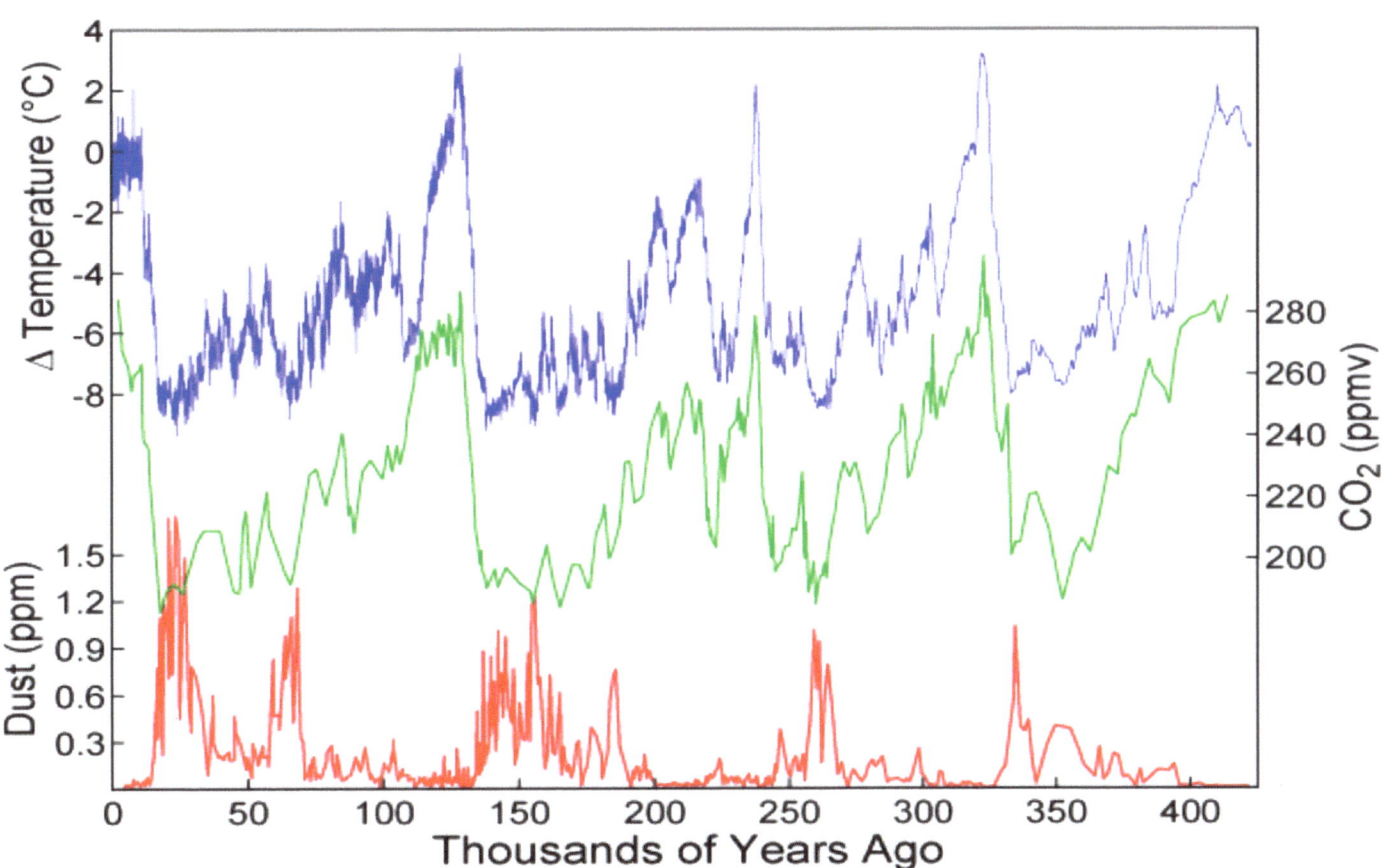

Chart courtesy of NOAA. (Apologies for the change in chart direction.)

Downloading the GISP2 data, for research into the dating of volcanic eruptions, I noticed something even more scary. The transition period from 'Warm' to 'Ice Age' conditions can happen quickly with little or no warning, but even worse, if history repeats itself, it could happen to us in as little as one thousand, but no longer than ten thousand years from now.

Searching for more information about this transition period, my attention was drawn to what the peat bogs in our northern tier of states could tell us. It turned out that the switch from 'Warm Periods' to 'Ice Ages' only took eight to ten years to take place. (Not the total time from peak to valley, but rather it was a straight and steep line downwards in temperature with the transition sudden and brutal.)

Next, came the discovery of the deepest ice sheet yet, at eleven thousand feet deep, it promised to provide data going back some 800,000 years (Dome C or EPICA ice core data). It was now that the alarm bells on climate change truly started to ring for the Greenland data had shown four 'Ice Ages' in 400,000 years, while this one showed eight 'Ice Ages' in 800,000 years. In other words, it was a cyclical event. Even more frightening was the fact that the 'Ice Ages' were much longer than the 'Warm Periods' accounting for 75% of the time. It was now 2008.

2008 Dome C Ice Core Data. (EPICA-C)

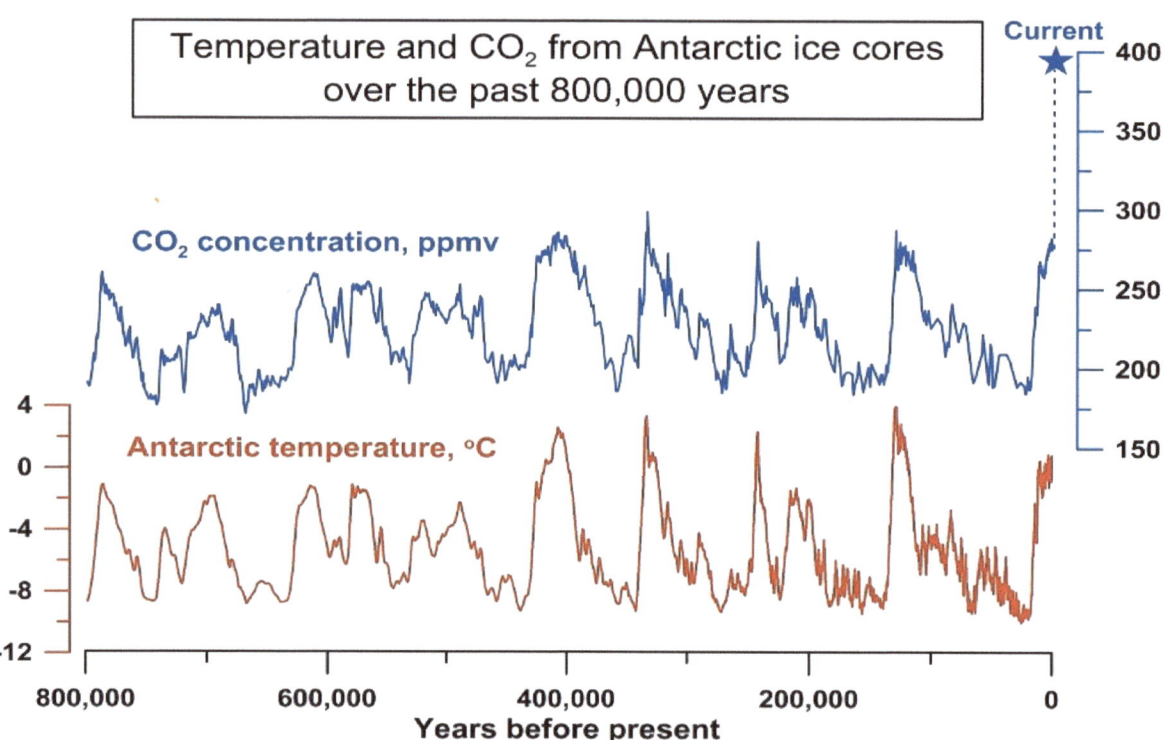

Chart Courtesy of Wikipedia/NASA. Please note the chart is back to normal with current data on the right.

Curious, I wanted to see how quickly the transition from warm to cold was taking and the charts I developed took an entire page to cover a few thousand years, but the following one still helps.

Volcanologists please note: the eruption of the Toba Volcano in 74,289 BP, with an ejecta volume of 2,800 cubic kilometers of tephra, left a SO_2 marker with a temperature drop that lasted until 73,507 BP or 732 years. I posit that only a plume reaching into outer space could have done that and would be the reason that scientists thought that SO_2 was from a biological source. In other words, it had a massive impact on climate, particularly in the geographic areas not affected by glaciation. As for the smaller Yellowstone eruption in 640,000 BP, that ejected 1,000 cubic kilometers of tephra, but it has recently been discovered that this was not one eruption but two, 170 years apart and each consequently smaller and with less of an impact on glaciation reinforcing the comments of Zelinsky[2], but noticeably extending the coldest period by a mere 5,000 years.

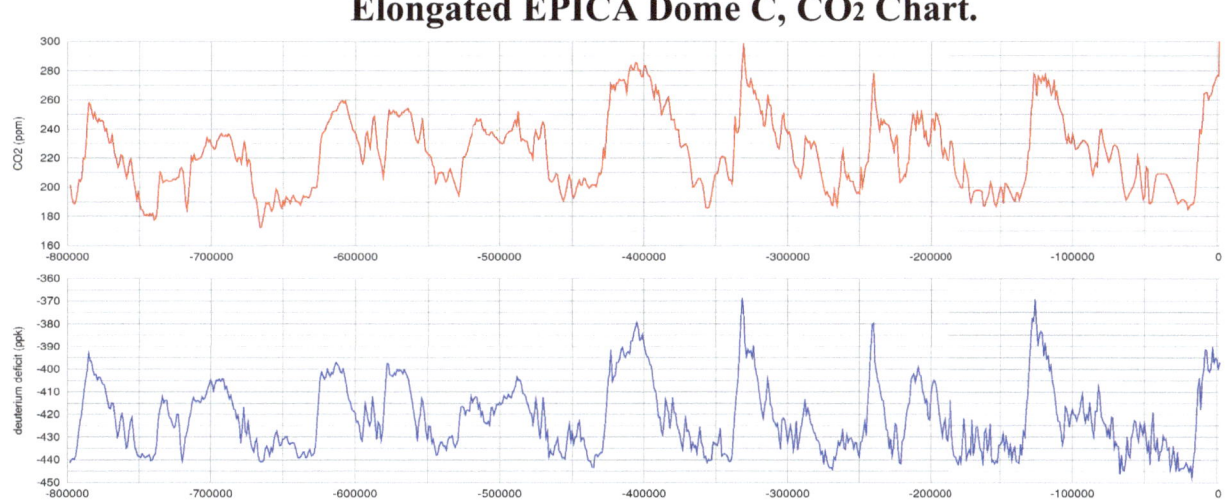

Elongated EPICA Dome C, CO_2 Chart.

With the dividing line between ice ages and warm periods around -4 °C, and the deuterium level at 415 (ppk), this means that over 800,000 years and 8 ice ages, 75% of the time our Earth is in an ice age and 25% of the time in a warm period. Even of greater concern is that we appear to be half way through this cycle with the downturn perhaps being signaled, but as yet from looking at history, uncertain. We needed more history. The other factor that showed up on these expanded charts was a fact that needs explanation from our scientists for temperature and carbon dioxide levels do not move in a synchronized manner but in fact both precede and follow one another with a lag in time of up to ten thousand years. ('Deep Ocean Overturning' and more on that later?)

Another factor this elongated chart brings out is that the warming period from 800,000 to 400,000 years ago was 2° C colder and implied a shorter and more modest warming period, something that is truly alarming for the survival of humanity. World scientists meanwhile had been searching for ways to extend our knowledge of climatic conditions even further back in time. The deep ocean proved to be the answer, for sediment deposits were undisturbed by wind and rain remaining untouched for millions of years. The clue to tracking climatic conditions was from the tiny shells of plankton (krill) that had died and sunk to the bottom of the sea. The calcium was the way they tracked temperature and a host of other information. First, they released data going back some 5 million years, and was compiled from 57 different cores. My wonder was why nobody was raising alarm bells about the fact that these older ice ages were colder to the point that the warm periods might have been so cool that the ice ages had blended into a two-million year-long freeze.

---

[2] Klarissa N Davis et al, Columbia River Basalt eruptions: WSU: Geology magazine.

Despite this new information, for me, it only proved that these ice ages went back 2.8 million years, and then inexplicably lessened, with the temperatures showing smaller variations. Becoming of even greater importance was the why with many theories being posited, but without convincing proof. It was now 2005.

Ocean Floor Drilling Records

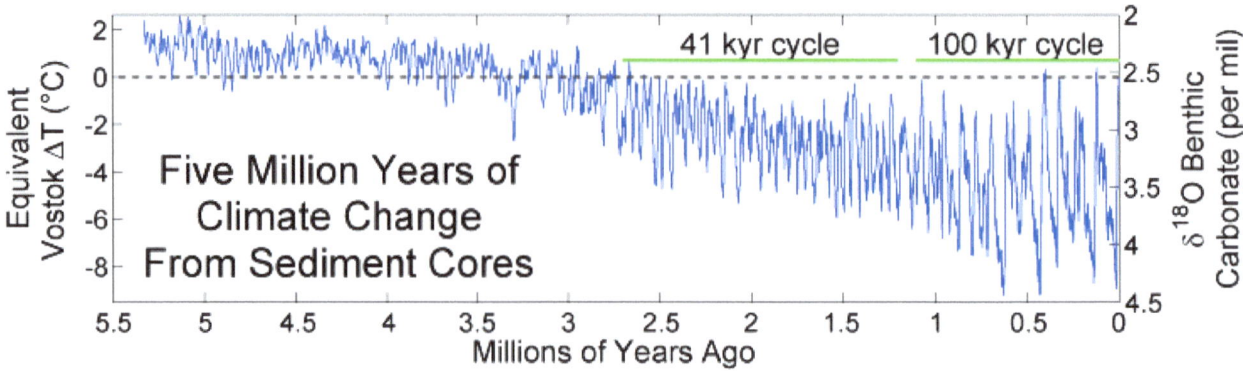

Chart courtesy of Wikipedia/ NOAA/Geoff. Zeiss.

Later, these ocean bed cores traced that history back to an incredible 65 million years. What they showed was a record of fluctuating 'Ice Ages' going back some 3 million years, and then a much smaller temperature fluctuation for another 12 million years, and then before that, None! Screaming to me was the question of why were these changes happening? Indeed, the relationship between carbon dioxide in the atmosphere and temperature seemed no longer to govern with other

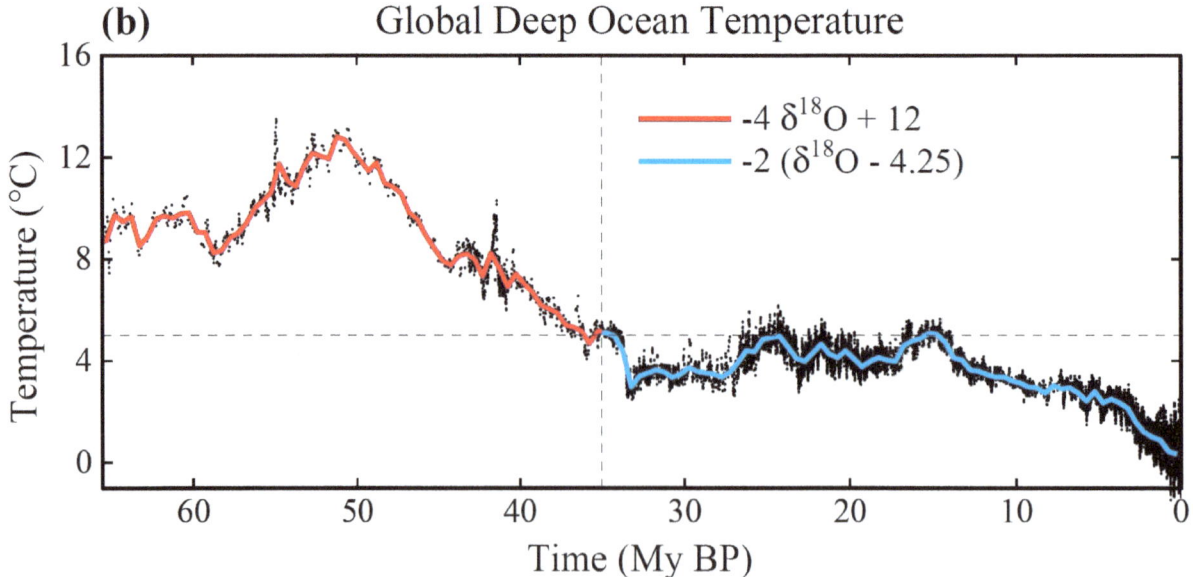

Chart Courtesy of Geoff Zeiss/NOAA.

conditions over-riding that relationship.

If I was skeptical before, I was now truly alarmed for Al Gore now presented his update on "Climate change". As before, it was about carbon dioxide and global warming with almost nothing about the threat of another 'Ice Age', nor of the escalating damage that was happening to the environment around us. A series of charts now appeared that helped combine the information we were seeing into a more understandable profile by combining them with the Earth's geological history and its major events. The charts again raised the question of what was causing these temperature changes? From 58 to 51 million years ago (mya) the temperature climbed 4°C, then dropped 4°C from there to 44 mya. From that point, with continuing fluctuations, from then onwards it cycled back and forth for the next 18 mya until 15 mya when it proceeded to drop another 4°C until 2.8 mya. Then, something strange must have happened, for not only did the temperature continue to drop, but the fluctuations in temperature started to widen, bringing us into the 'Ice Age' cycles recorded in our Antarctic and Greenland ice core charts. It was now mid 2017.

The Combination of all Temperature Data.

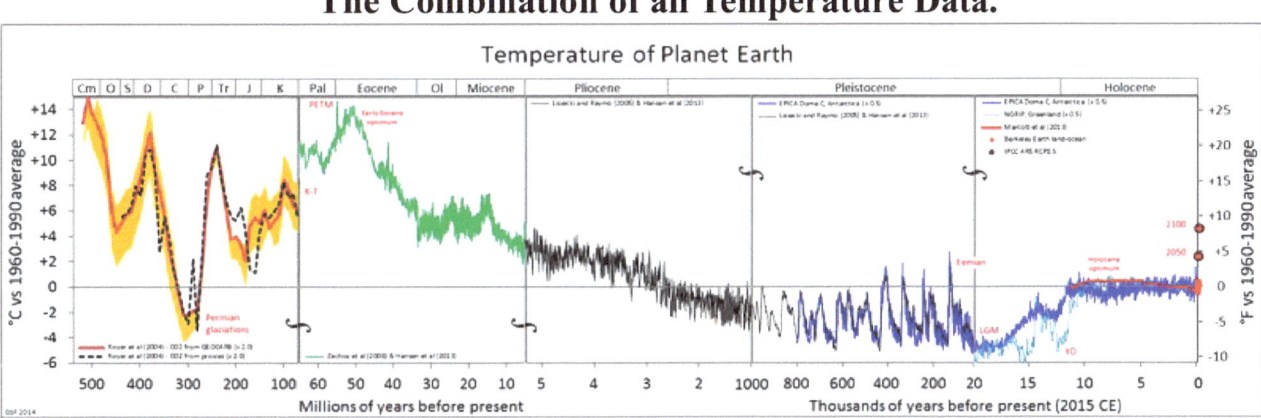

All Charts courtesy of Wikipedia and NOAA.

The Combination of Temperature and Geological Data.

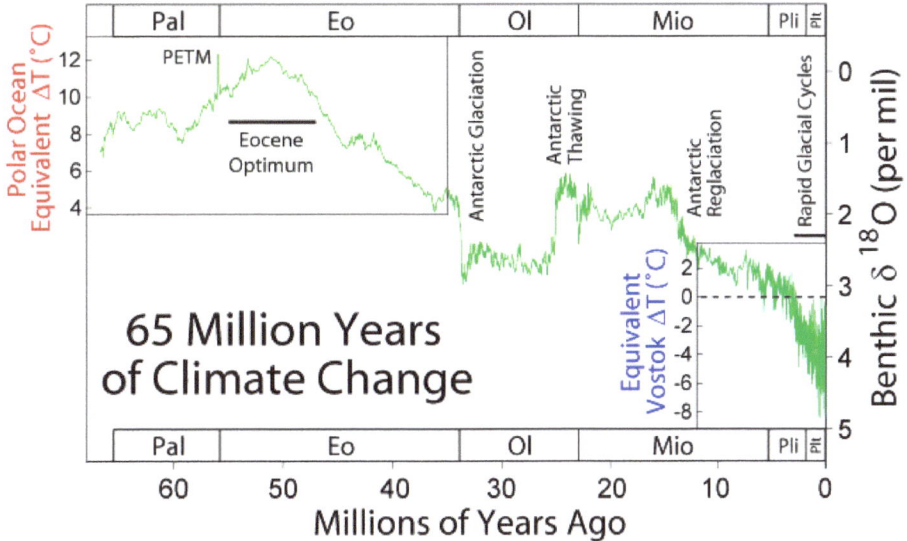

And now to a subject I hate, for it masks the fantastic work our scientists have been doing. 'Bad Science' is the trap for all researchers and extrapolating short periods of correlated factors into millions of years was the trap. The flaw was the fact there were several other more dominant factors that came into play and were responsible for setting World temperatures and so extrapolation has its limits, something that scientists have long known.

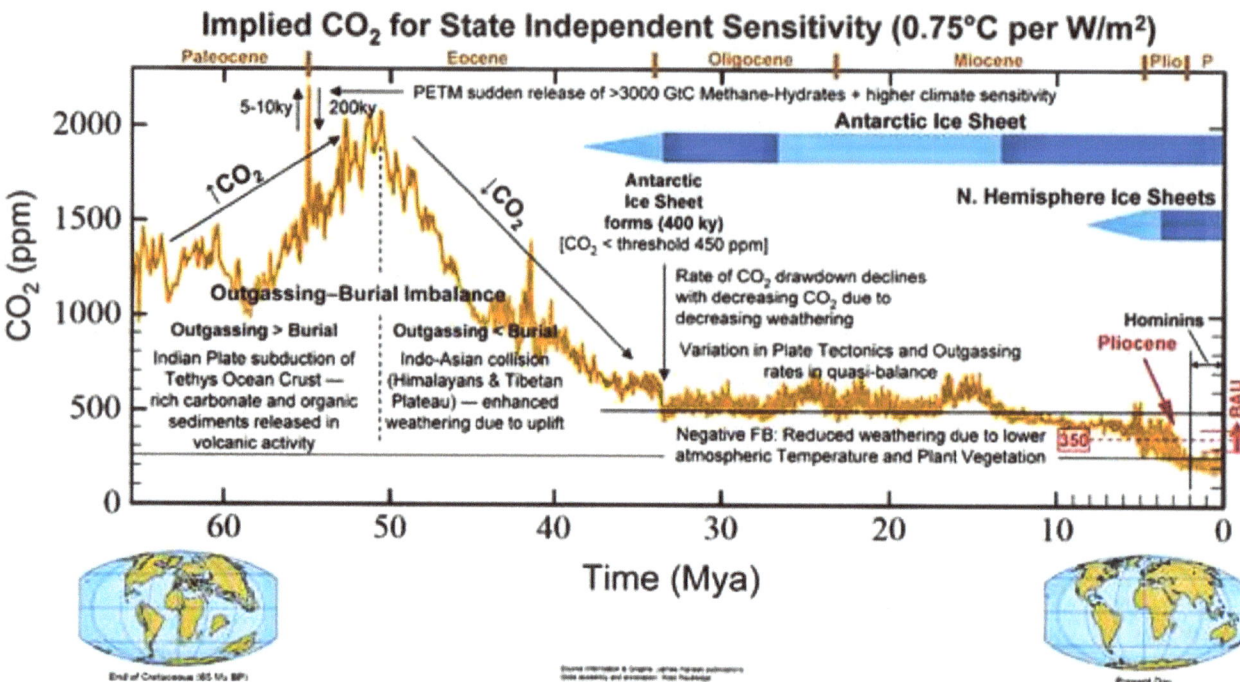

Perhaps all was not lost for now we could compare the hypothetical to the actual recorded numbers in the following chart.

The Paleomap Projection.

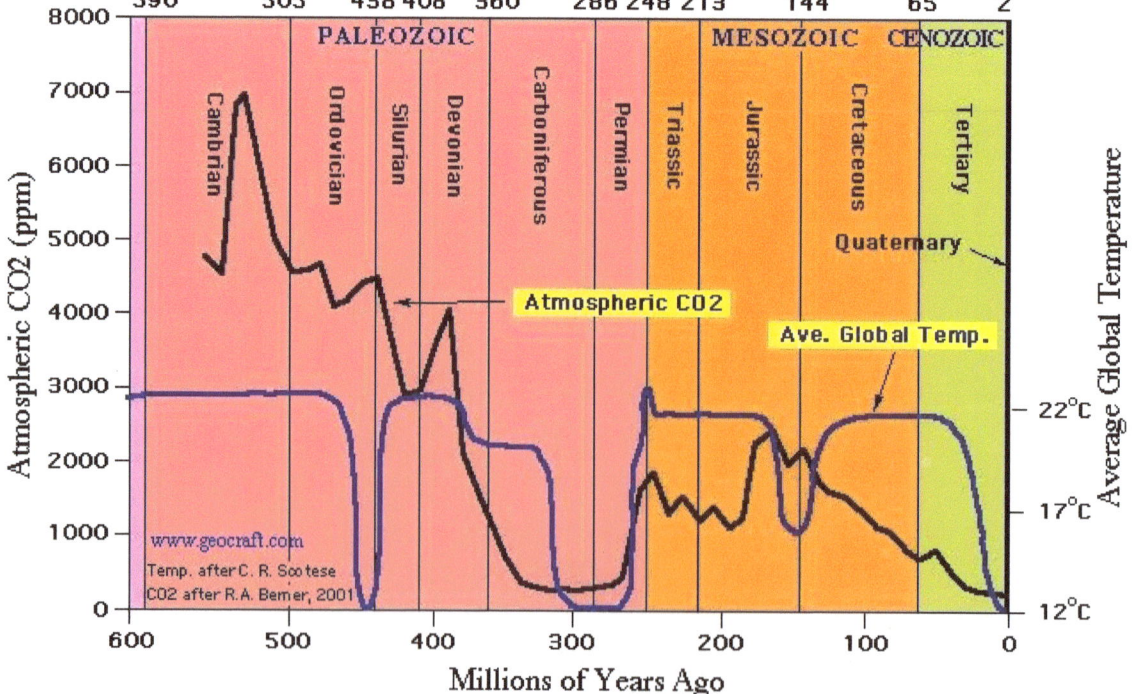

Image courtesy of C. R. Scotese, PaleoMap project. (Please note that the mya temperature has been adjusted with the terminology of average global temperature correct!)

Deep Ocean Temperature Range in million years (Data in degrees Celsius).

Years mya	67-45	45-35	35-5	5-0	Total Change 67-0
High	12	7	2	-4	-16
Low	7	3	-3	-11	-18
Avg.	9	4.5	0	-7	-16
Degrees F	48.2	40.1	32	19.4	-28.8

(All data from NOAA National Climate Data Center, courtesy Carrie Morrill.)

If one considers that today we are locked into a cycle of ice ages, the earlier period before three million years ago looks infinitely preferable versus today's, with the more distant past ones survivable but frightening. The question for our experts is, if we can impact weather by our carbon dioxide emissions, why would we not strive to raise them to a number say half way between our recent past or about 4C (7.4F) degrees? In case anyone has not noticed, the Paris Climate Accord set a target of 2C (3.7F) degrees which by the time the change to renewables is completed may be closer to the above range and might avoid the next ice age, a perfect solution. And that is why we should endorse it as soon as possible with only one change. Make sure the USA does not have to hand over all that money to other countries to make it happen. China, Korea, Japan, and Europe should share that burden.

Does this mean that 800 ppm of CO_2 should be our target and not some lower number that would kick us into the next ice age? The idea sounds outrageous, but might make sense even if it does not solve the issue of the physical damage being done to our earth, and so for that answer, we need to continue.

For anyone who would like to explore the many variables involved in this analysis, I consider the work of Professor James Zachos at UC Santa Cruz to be one of the most brilliant I have read in recent years. Rest assured I will be sending him a copy of my book in the hopes he can give us a definitive answer.

What are the climatic factors that were so great that they over-rode the short-term correlation between carbon dioxide atmospheric levels and temperature?

The first was the location of our Continents (continental drift) and what that did to our ocean currents and climate.

The second was the part volcanic activity played in determining the level of carbon dioxide in our atmosphere.

The third was the role that our Oceans played in acting as a storage or balancing system for that carbon dioxide.

The fourth was how that impacted on our carbon cycle and the gradual shift from the amount in our short versus long-term carbon storage deposits (deep Earth).

The fifth was the burning of fossil fuels in quantities no one could have predicted, nor could they have forecast the impact it would have on our short-term carbon cycle.

Here, the astronomers came forward with one of the greatest. The Sun's age and how its heat emission was changing. Secondly, but of relatively minor importance, the position and oscillation of our Planet in relation to our Sun. Further, in the period in which we are most interested, 65 mya until now, that will not to our knowledge have played much of a part.

For myself, I wanted to find out what caused these huge changes in temperature, the start of ice build-up, the cyclic swings between warm and cold periods, and finally the even wider swings that led us into the age of repeating 'Ice Ages' with the final question of what happened 1.2 mya that put us into the current period of even greater swings in climate.

Already thinking I knew the answer, the sophistication of the recent research by our geologists was breath-taking. Using cutting edge measurement devices, they have brought a new level of understanding to that science.

*****

4. CONTINENTAL DRIFT.

New technology is allowing our Geologists to rewrite the story of 'Continental Drift'. However, when talking about the movement of Continents, one has to think in millions of years and recognize that this movement is happening at a snail's pace. Likewise, Ocean current and climates change gradually in concert with these movements. Sometimes though, especially when Continents collide and open or close new passageways between our Oceans, climate change can be abrupt and dramatic. The data being gathered is so sophisticated that our geologists have been able to identify at least two if not three previous 'Continental Drift Cycles' that are even today playing an important role in the movement of our Continents. As an example, our own Rocky Mountains were formed by an earlier subduction event and explains why they are so far inland from our current west coast subduction zone. Likewise, these ground penetrating instruments have identified that the west coast of S. America has broken off and is being currently sub-ducted as that continent drifts westward. While these are just interesting examples of what is being discovered, that growing knowledge might revolutionize our understanding of how our Earth behaves when 'Ice Ages' occur and the 'Rebound Effect' those massive ice sheets have on the surrounding areas, the sea-level changes, ocean currents, and of course, climate.

Continent Positions 60 million years ago. (mya)

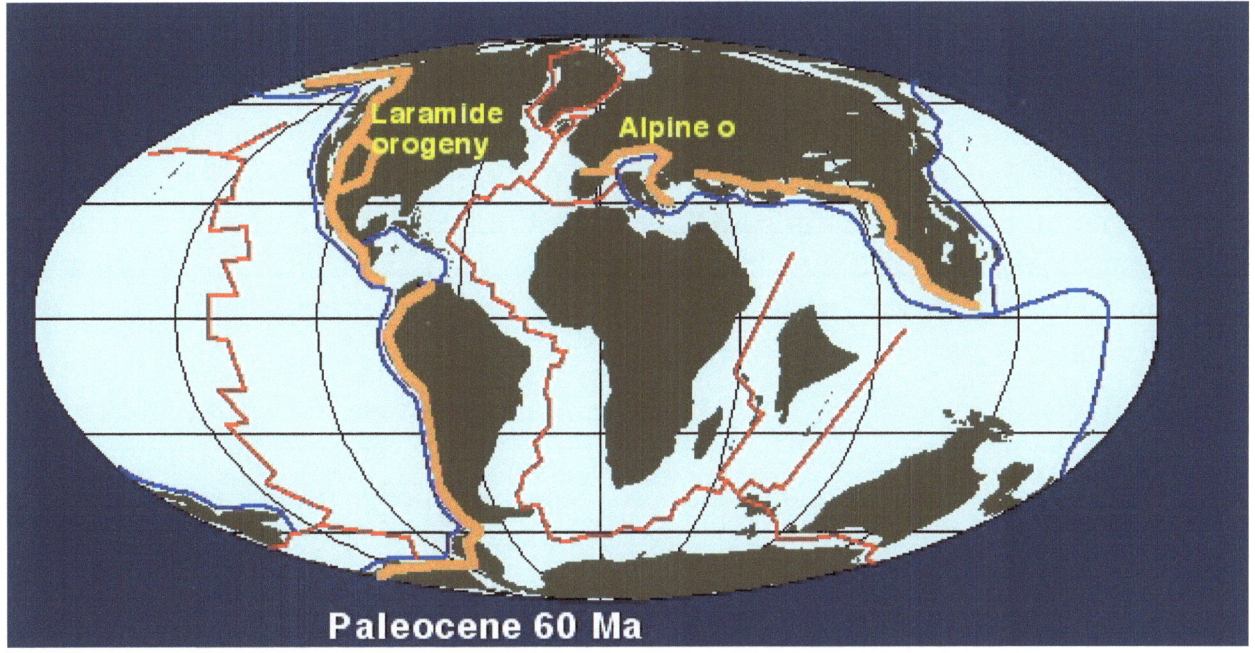

These six charts are courtesy of C.R. Scotese,1998, Quicktime Computer Animations, PALEOMAP Project, Dept. of Geology, University of Texas, Arlington, Texas. Dr. Ron Blakely, Professor of Geology, Northern Arizona University. The tectonic data are derived primarily from Scotese and Glonka, 1992, Paleomap, U of T, Arlington, Texas: Ziegler, 1998, Evolution of the

Arctic-North Atlantic and the Western Tethys, AAPG Memoir 43

S. America was attached to Antarctica and N. America attached to Greenland and Europe. The two America's were still separate as was India. However, that Sub-Continent was no longer over the Reunion Island volcanic 'hot spot' and the Deccan Traps had ceased flowing, reducing its fantastic outpouring of carbon and sulfur dioxides into the atmosphere.

'Continental Drift', or how the land masses were constantly moving around our globe has been the key to answering many of our questions. Starting with what happened 58 mya, it was a rare event in history for not the collision, but the speed with which the Sub-Continent of India was moving. Though separate and perhaps being pushed by the Indian plate, was another one underlying the ancient Tethys Ocean (that part long ago sub-ducted under what is now Tibet). Crashing into the almost immovable Eurasian plate, the Tethys plate was driven under it with the softer sediments on top being scraped off and forming the sedimentary rock layers which formed the Himalayan Mountain Range. With that scraped off layer being half a mile deep and the plate moving at the rate of nearly 240 miles in one million years, this was a 40 or more million-year building project and explains why the actual collision of the 75-mile deep Indian plate did not occur until 20-15 million years ago with subduction eventually the result. In terms of climate, this ever-growing colossal range of mountains played a huge role in reducing the 2,000-ppm level of carbon dioxide in our atmosphere, and an even higher level of sulfur dioxide (resulting in highly acidic rain), in that it formed a barrier, with monsoon-like downpours falling on the Range's southern flanks where the acid rain reacted violently with the sedimentary material that makes up the Himalayas. Another factor in play was the deep plume located under the island of Reunion where the Sub-Continent of India was located at that time. Highly active from 67 to 66 mya, when the 'Deccan Traps of India' were laid down in a series of layers, this was a truly vast volcanic event laying down an over 6,700 foot deep layer of lava (basalt rock) over a 600,000 square-mile area, or half of India. It was probably no coincidence that the speed of drift reached its peak between 67 and 63 (mya), when it then started to slow down, for at that time it was over the hot-spot that is currently under Reunion Island.

Geologists have theorized that global temperatures rose 4° C from 58 to 51 mya due to the creation of the Himalayan Mountains blocking off air currents. They then proposed that the monsoon rains that developed washed out the excessive carbon dioxide from the atmosphere (from the Deccan Trap formation) in the period 51 to 44 mya with the resultant acid rain reacting quickly with the calcium in the rocks that formed the Himalayas explaining both the rise and fall of temperatures and carbon dioxide levels at that time. Also occurring in that period was the separation of S. America from Antarctica with the divide shallow with only the warm surface water moving from the Pacific into the Atlantic.

Continent Positions 40 mya

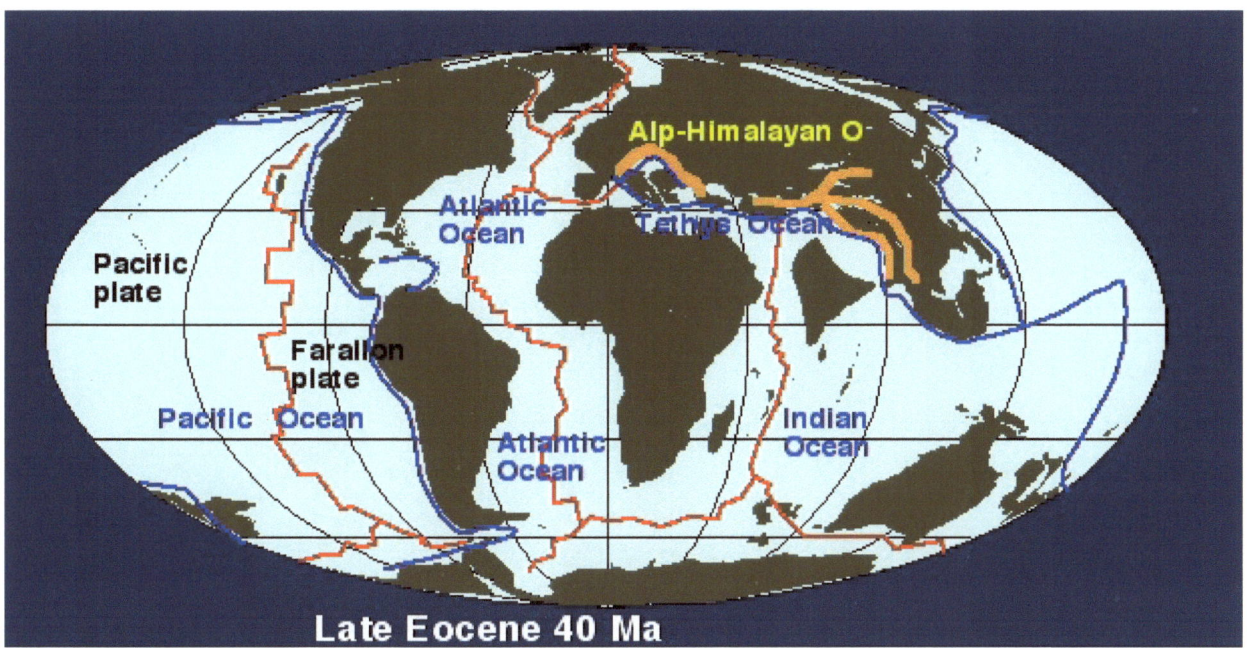

A gap opened up in two stages between S. America and Antarctica. The first allowed shallow and warm water to pass through from the Pacific to the Atlantic Ocean. The second was the opening of the 'Drake Passage' allowing deeper and colder water to flow through.

Continent Positions 30 mya

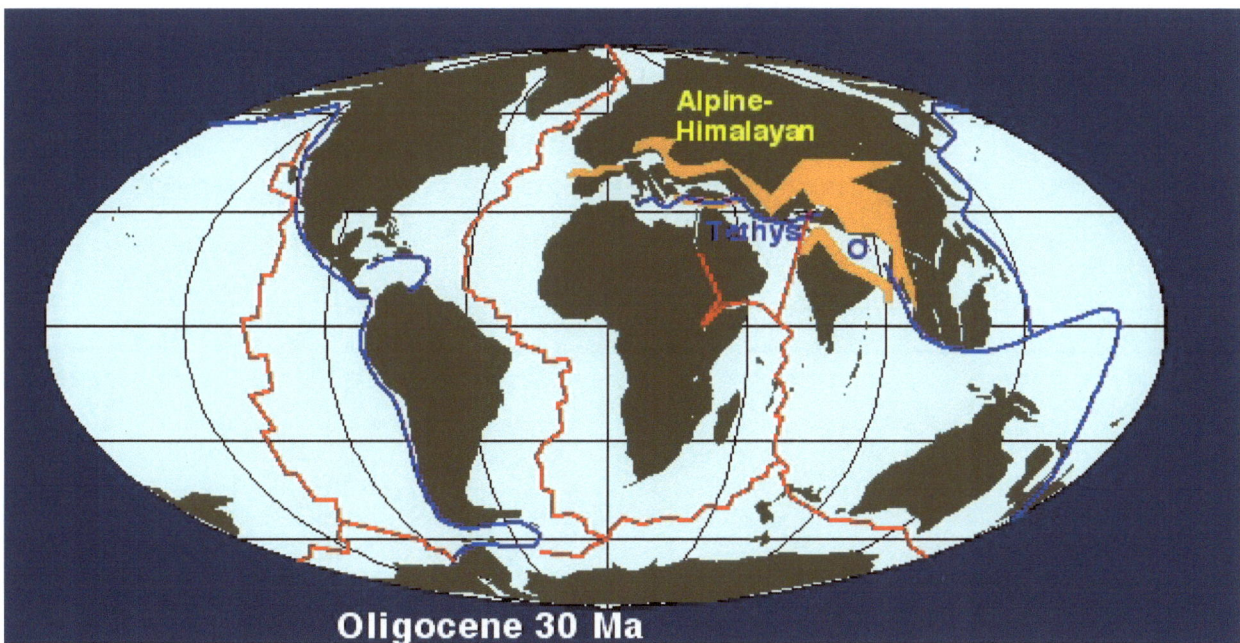

The Ocean plate of the Tethys Sea was sub-ducting under the Eurasian Continent, and with the softer ocean deposits being scraped off as it sub-ducted under the Eurasian plate, forming the Himalayan Mountain Range with dramatic changes to the area's climate.

Continent Positions 20 mya

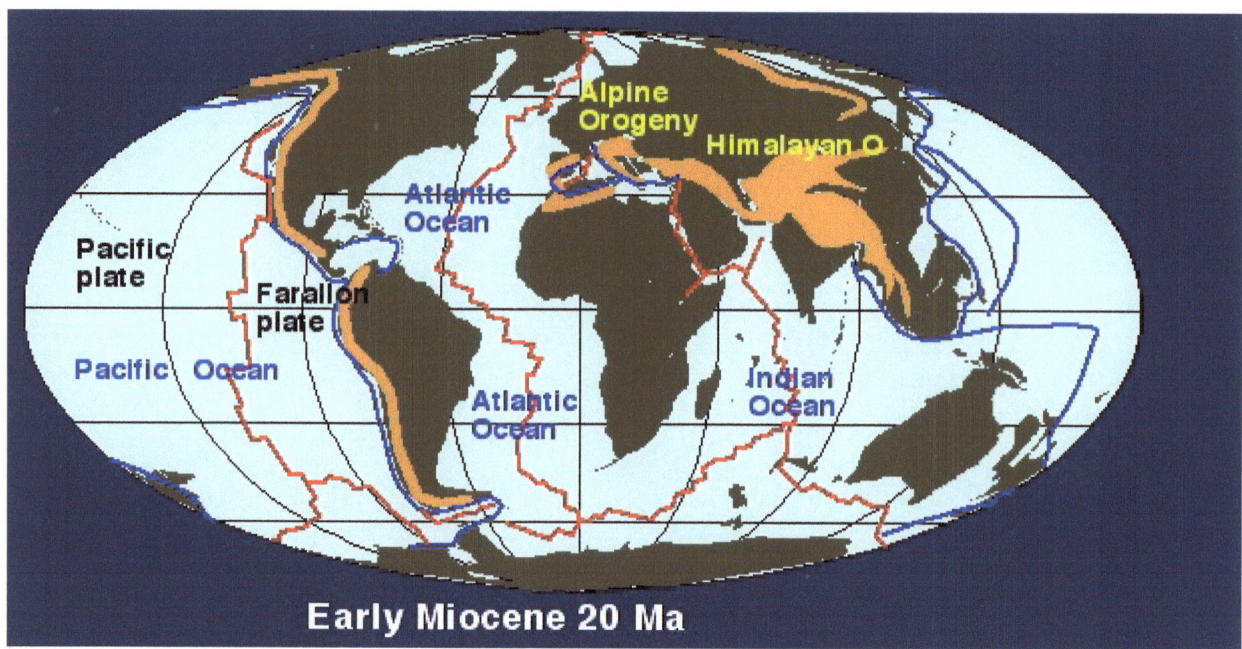

The Sub-Continent of India finally collided with Eurasia. By this time N. America had finally broken free of the ocean ridge connecting Greenland with Europe and with the gap between N. and S. America narrowing and the gap between S. America and Antarctica widening. The main deep-water ocean current between the Pacific and the Atlantic Oceans now moved South creating the first permanent Ice sheets in Antarctica with cyclical variations appearing for the first time. Also occurring was the closure of the Tethys Ocean between the Indian and Atlantic Oceans forming a dry Mediterranean Valley.

Continent Positions 10 mya

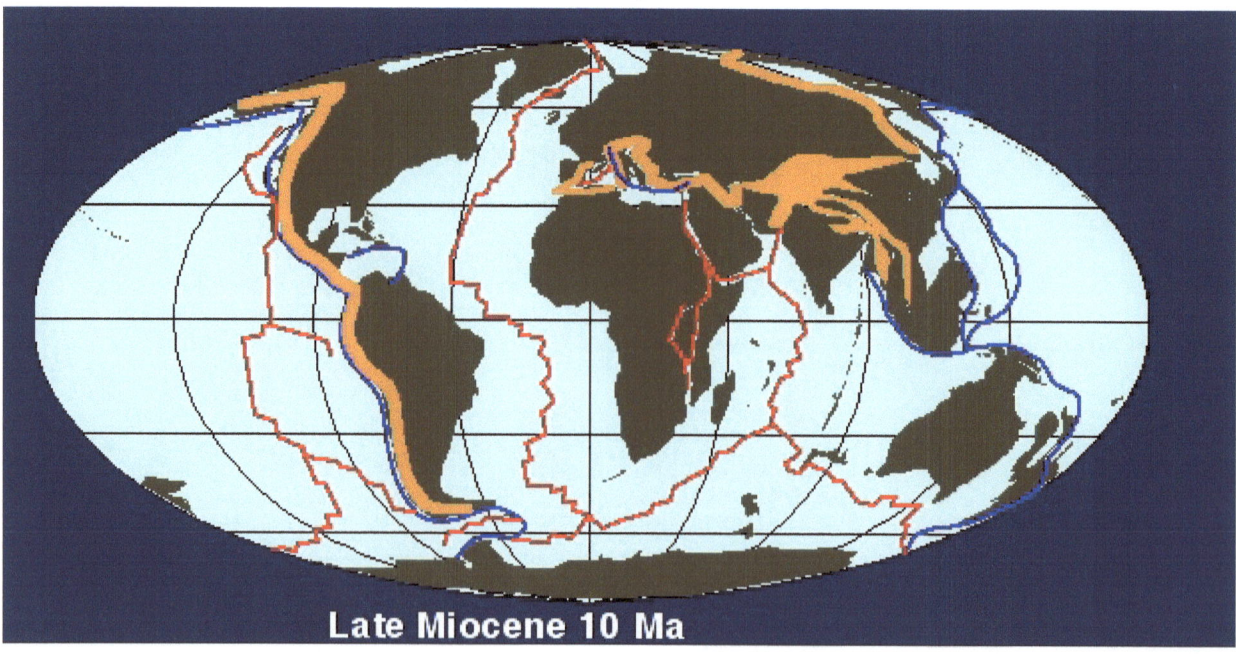

With only warm water flowing from the Pacific to the Atlantic Ocean between the two Americas, and with the Arctic Ocean now Connected to the Atlantic deep ocean currents, the variation between cold and warm periods widened.

Continent Positions Present Day.

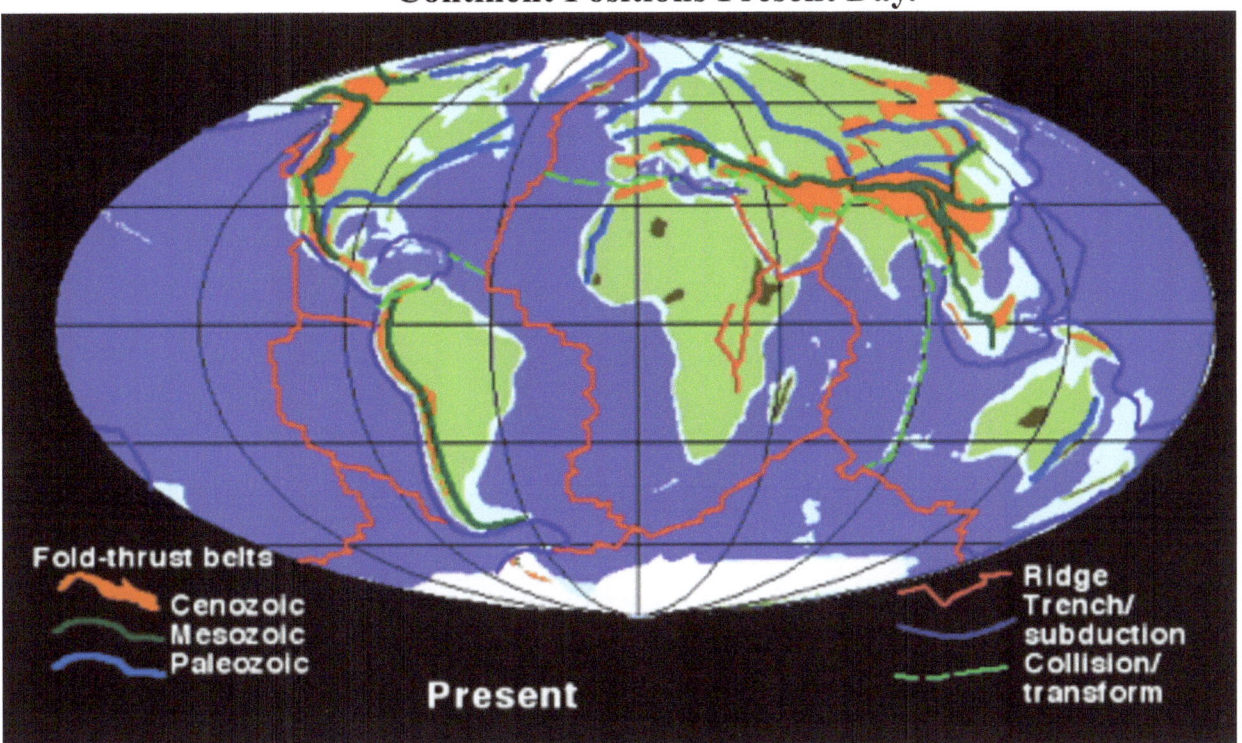

With the gap between the two Americas now completely closed by 3.5 mya, the variations between cold and warm periods widened dramatically ushering our Planet into a pattern of 'Warm and Ice Age' cycles disclosed by the 'Ice Core' drilling projects. (Please note that the Ocean level drops that resulted from Ice Level build-up and the rebound effects that resulted from melting, shut off the ocean flow from the Pacific to the Arctic Ocean through the Bering Strait, causing another disruption in the Planet's 'Climate Balancing System'.

The deep ocean and colder water current did not develop until the opening of the Drake Passage in 34 mya which also coincided with a sharp 2° C drop in temperature over the next million years as that passage widened; it narrowed again in 25 to 15 mya.

Over the next 18 million years, two events occurred. First, the African Continent, moving northward, bumped into the European plate shutting off the Tethys ocean current to the Atlantic 13 mya.

Playing the main role in climate change were four huge events: the collision of the Indian sub-continent with the Asian plate, the separation of the South American plate from the Antarctica plate, the separation of the North American plate from the European plate, and lastly, the gradual joining of the two American plates at today's country of Panama.

The Continental Drift Summary Versus Temperature Change.

Mya*	EVENT (though not recorded, high levels of SO₂, largely off-set much of the CO₂ impact on temperature).	CO₂ ppm	°C
70-58	Deccan Traps, Comet strike and forest burning	3000	-2
58-51	Himalayas Mountain forming. / S. America Antarctic separating with warm water entering the Atlantic and Arctic Oceans.	2000	+4
51-44	CO₂ removal in acid rain/rock capture in Himalayas.	1500	-4
44-33	Continued CO₂ removal in Himalayas. / Cold deep water entering the Atlantic from the Pacific in 35 mya.	1000	-5
33-27	Greenland ocean ridge separates from Norway in 27 mya joining the Arctic and Atlantic oceans with fluctuations in Temperature thereafter.	750	-2
27-23	Variations in Temp. as Drake passage narrows and reopens.	500	-1
23-15	Mediterranean warm current starts to gradually close.	400	+1
15-7	Mediterranean warm water gap closes with western exit closed by 6 mya.	300	
7-3	Panama deep water gap closes.	250	-1
3-2.8	Panama warm water gap closes with Temp. variations widening to current levels.	250	-1

Please bear in mind that some of these events had transition periods that took millions of years to occur. *Mya (Million years ago).

*****

5. THE HISTORY OF WIND AND OCEAN CURRENTS

Obvious now is the fact that the level of our Oceans and the depth of the passageways between our Continents was going to be all important in estimating the climatic impact of our Ocean currents. In addition, I wondered if that might offer some clues as to what were the triggers that had set off the changes from warm periods to Ice Ages.

The topography below our Oceans dictates where cold and dense water wants to settle and indeed flow and so we first looked at what that tells us. There is another truism that we need to remember and that is; as temperature rises so too do ocean currents slow. Conversely, as temperatures fall; so too do ocean currents speed up. However, nothing is that simple for wind speed also reacts to temperature gradient and in turn has an impact on ocean current speed.

Wind Direction Combined with Upper Air Circulation.

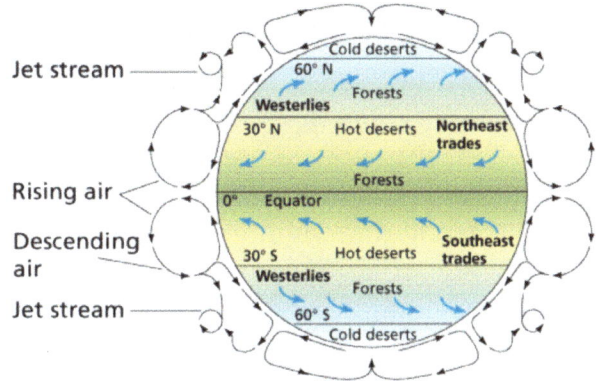

Image courtesy of Dr. Joseph Vincent Siri.

Surface Wind Direction Only

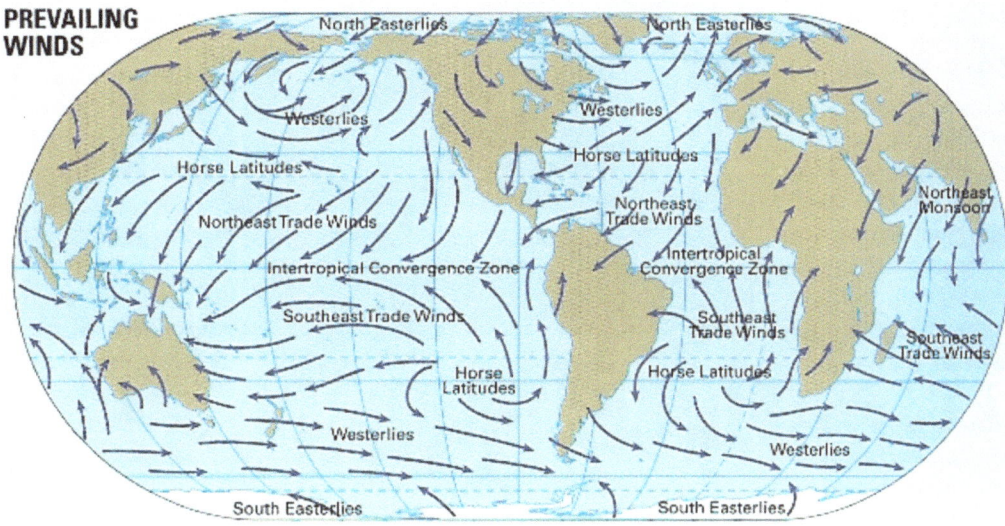

Courtesy of Unit Maps.

Not by coincidence is the fact that the strongest ocean currents match the direction of the air currents. One of the few exceptions is found in the NW Atlantic where the ocean current flows North to Iceland, and there it bucks up against the strong trade winds from the North.

World Ocean Topography

The Mid-Atlantic and Caribbean Ocean Topography

Images Courtesy of Wikipedia.

The subfloor picture of the North Atlantic shows that the Continents of North America and Europe must have separated in the distant past with that event dating back to around 60 mya when the last physical attachment to Scotland was broken. However, that was not the end, for to the North there was something else happening. Compression ridges were being formed as the North American Continent spun away from the European plate in the South with those same compression ridges forming a land connection between Greenland and Norway[3]. That detailed work showed that land bridge lasting to between 33 and 25 mya, a far different story from what had been the prior assumption. From this point forward, the width and depth of the gaps dictated where the deep ocean currents would flow. (Containing the densest and saltiest waters)

---

[3] Geological Society, London 2/20/14 The Faroe-Shetland basin. David Ellis & Martyn S Stoker.

The Present-Day Antarctic Ocean Floor.

The Arctic Ocean Topography.

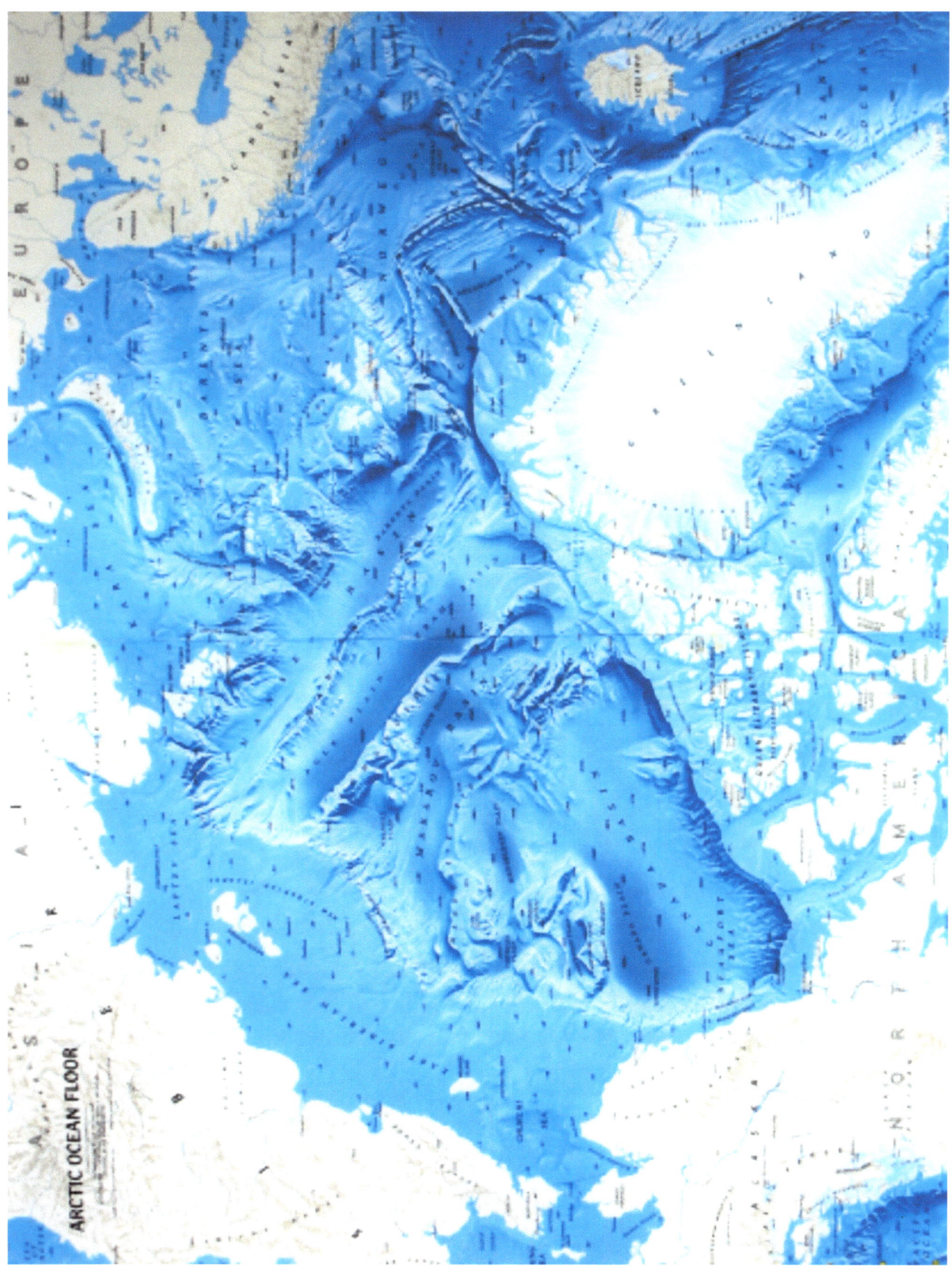

The South America-Antarctica Separation

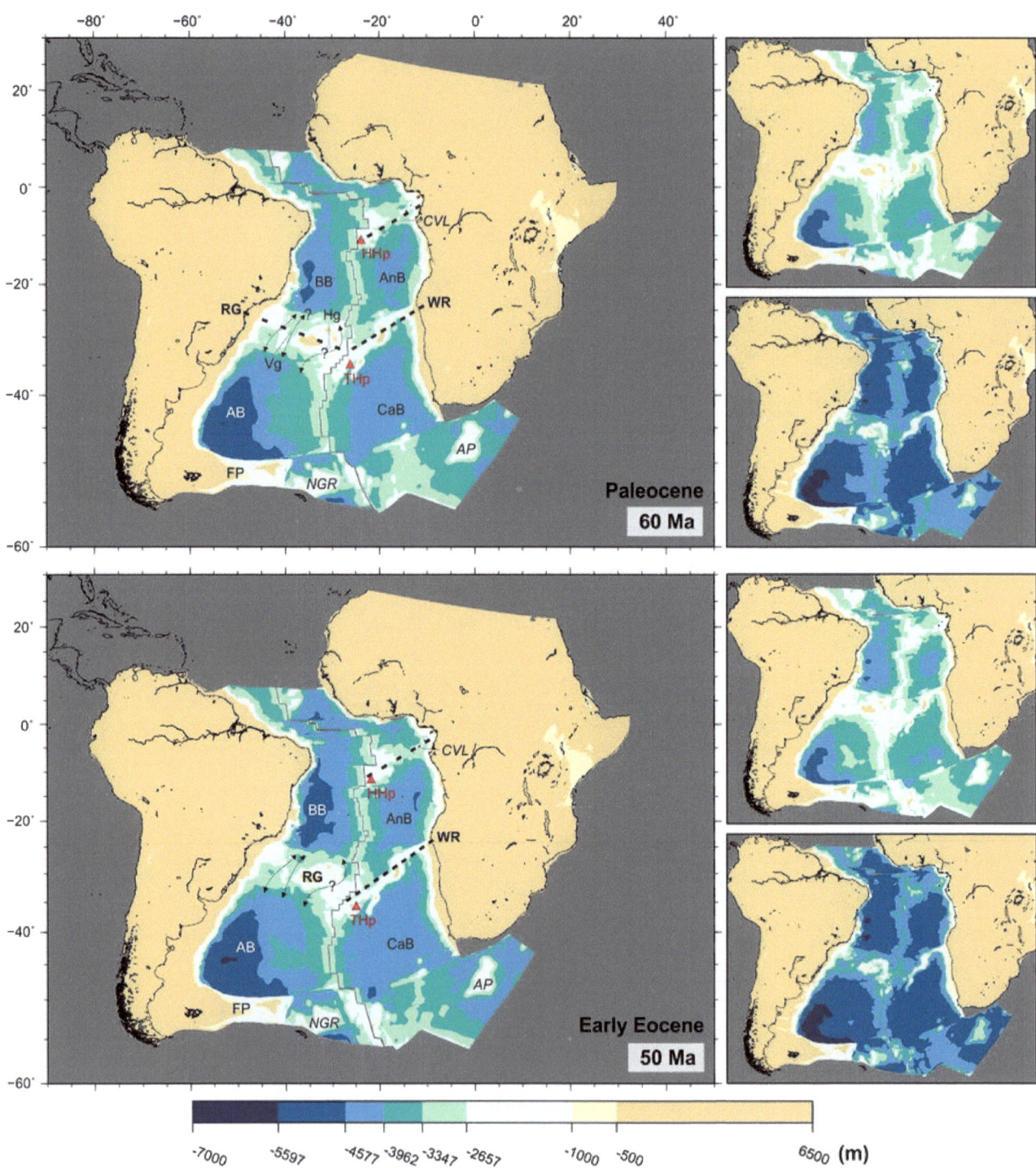

Courtesy PaleoMap Project.

The shallow current opening is estimated to have occurred in 60 mya whereas the deep current was estimated to have occurred in 35 mya. With each taking millions of years to widen, climate impact would have been gradual. The above map of today's ocean floor shows the deep 'Drake Passage' between South America and Antarctica and of course the huge passage between the Atlantic and Pacific

Oceans. Less obvious is what happened prior to those two Continents separation when the shallow spur was the only connection. The deep ocean current from the Pacific moves from West to East into the Antarctic Ocean, then circles around Antarctica in a clockwise direction, where it finally ends up spinning off Northwards into the Eastern Atlantic and then up to Iceland, where it plunges under the lighter fresh water flowing South from the Arctic. The following Maps show how the Atlantic and Pacific Ocean waters meet in this intermix in this somewhat complicated mixing bowl, but vital to understanding of how ice sheets can come and go in this part of the World.

Atlantic Surface Water Current in the Arctic Ocean.

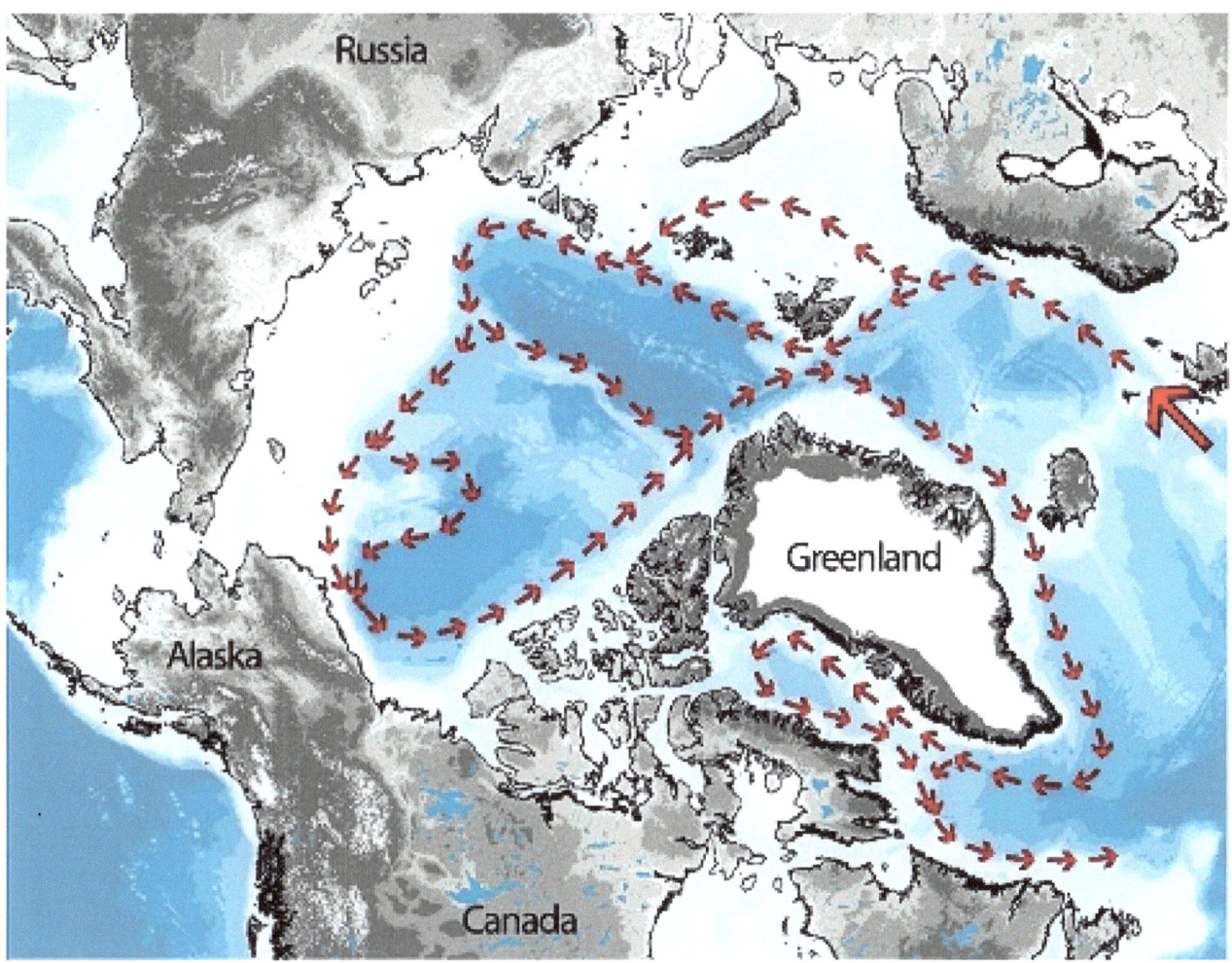

Map Courtesy of Govt. of Canada; 2016 Canada-Sweden Polar Expedition.

Please note that when ocean waters freeze, the salt condenses and drops to the bottom of the ocean with the resultant lighter ice and water staying on top. Conversely the water coming in from both the Atlantic and Pacific is saltier and seeks the deeper ocean where it can drop below the lighter water which dictates the current in that area.

Earth, a Planet in Peril

Pacific Surface Water Current in the Arctic Ocean

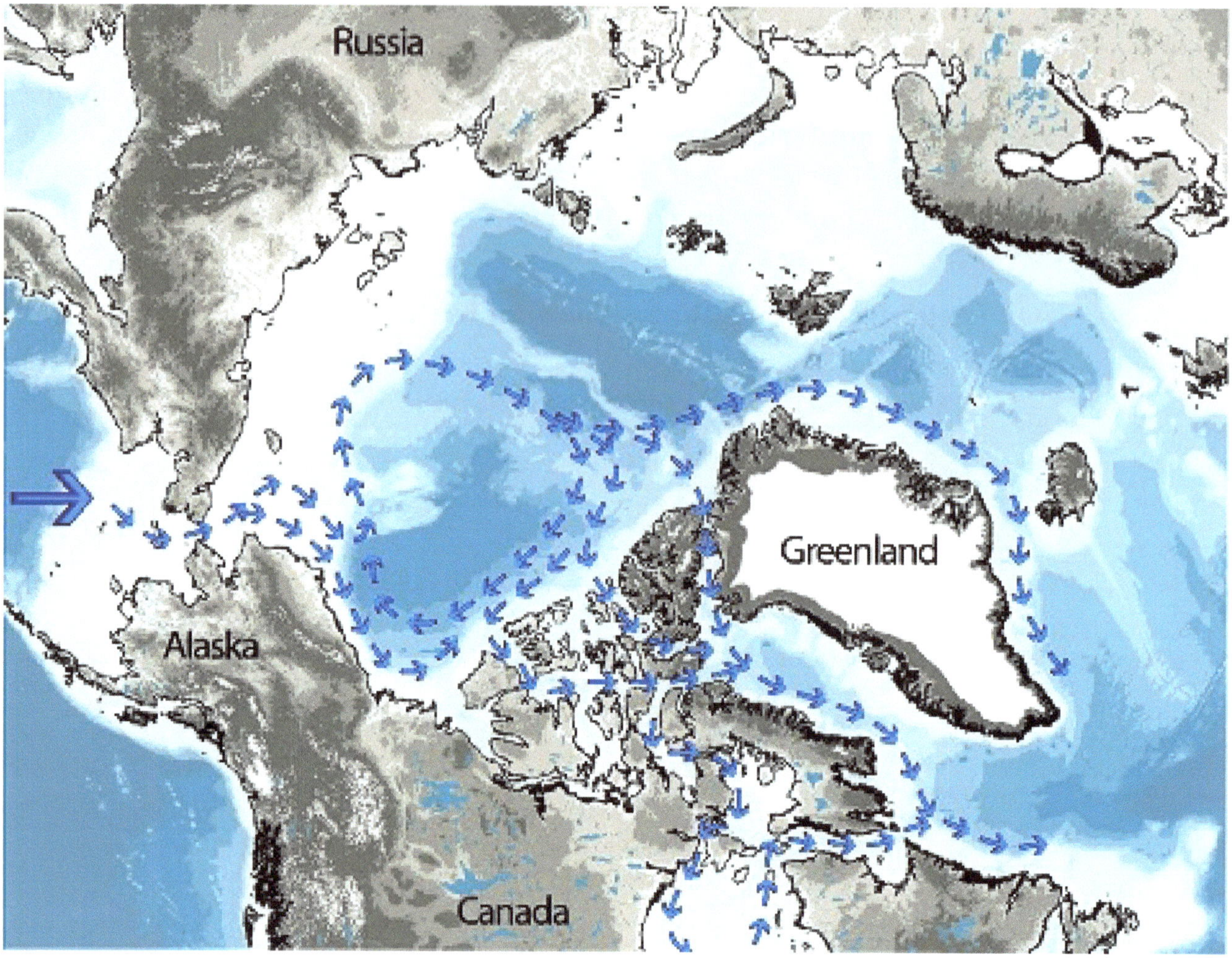

Map Courtesy of Govt. of Canada; 2016 Canada-Sweden Polar Expedition.

Profile of Present Arctic Ocean Depths and Water Layers

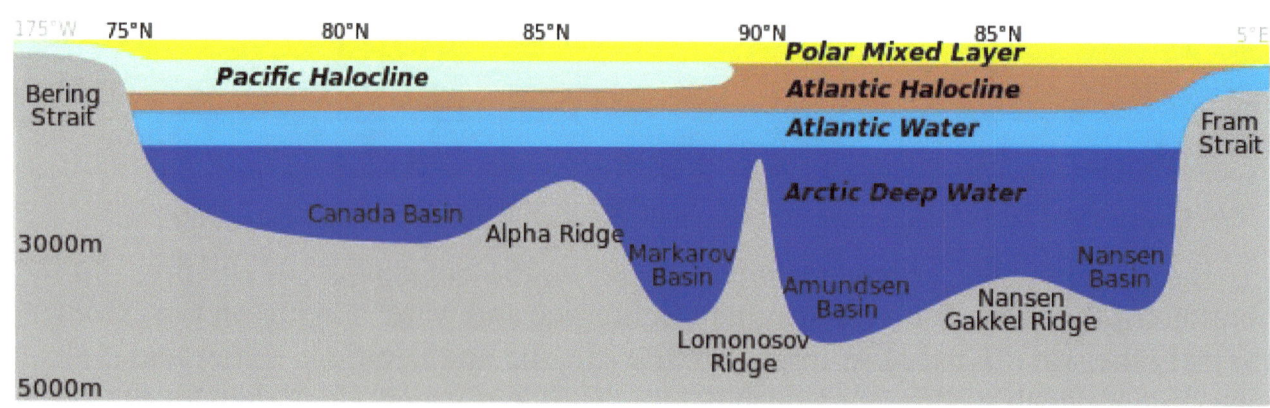

Map courtesy of Govt. of Canada; 2016 Canada-Sweden Polar Expedition.

Canadian Arctic Islands NE of Greenland.

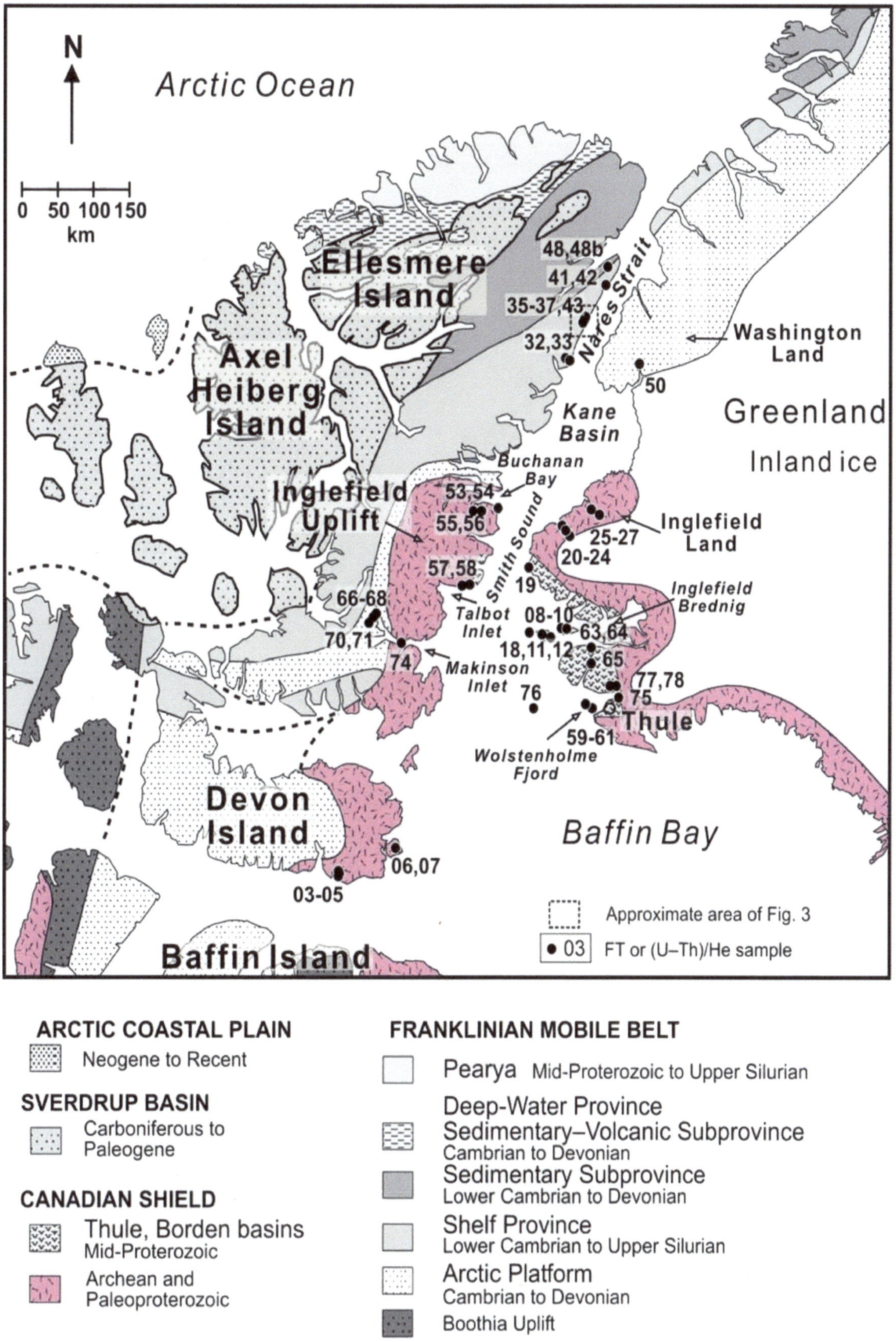

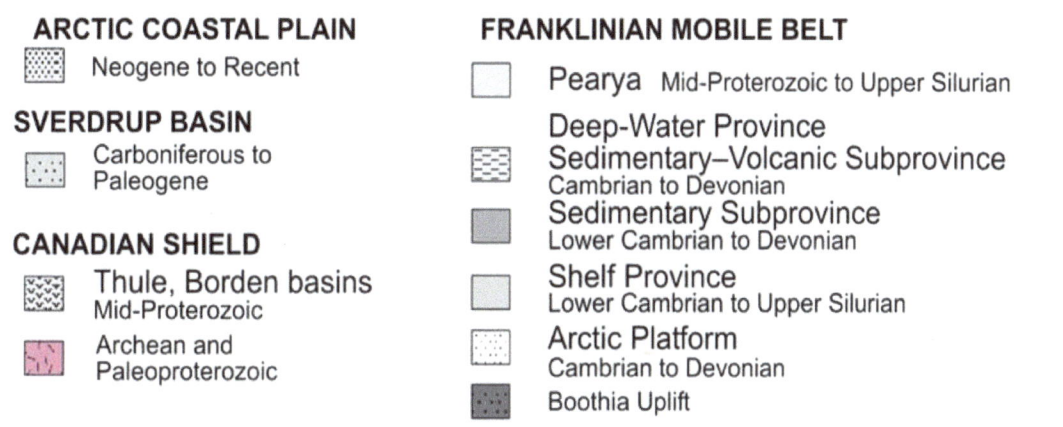

Earth, a Planet in Peril

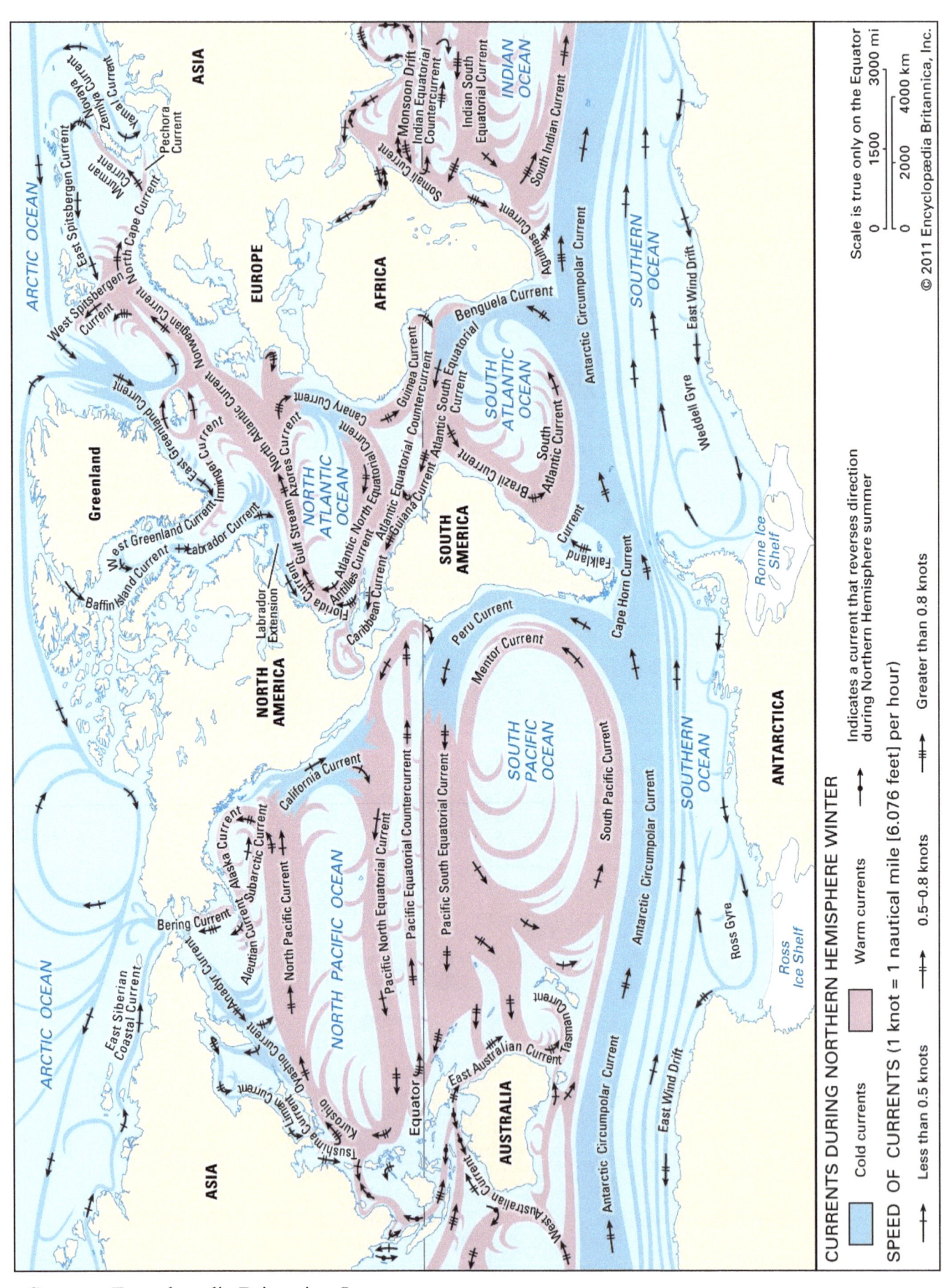

Courtesy Encyclopedia Britannica, Inc.

The above map of winter ocean currents as the best in that it has been scaled to also represent the volumes of water being moved, and thus showing the importance of the many currents in terms of their temperature balancing role.

The Arctic Ocean Current Program.

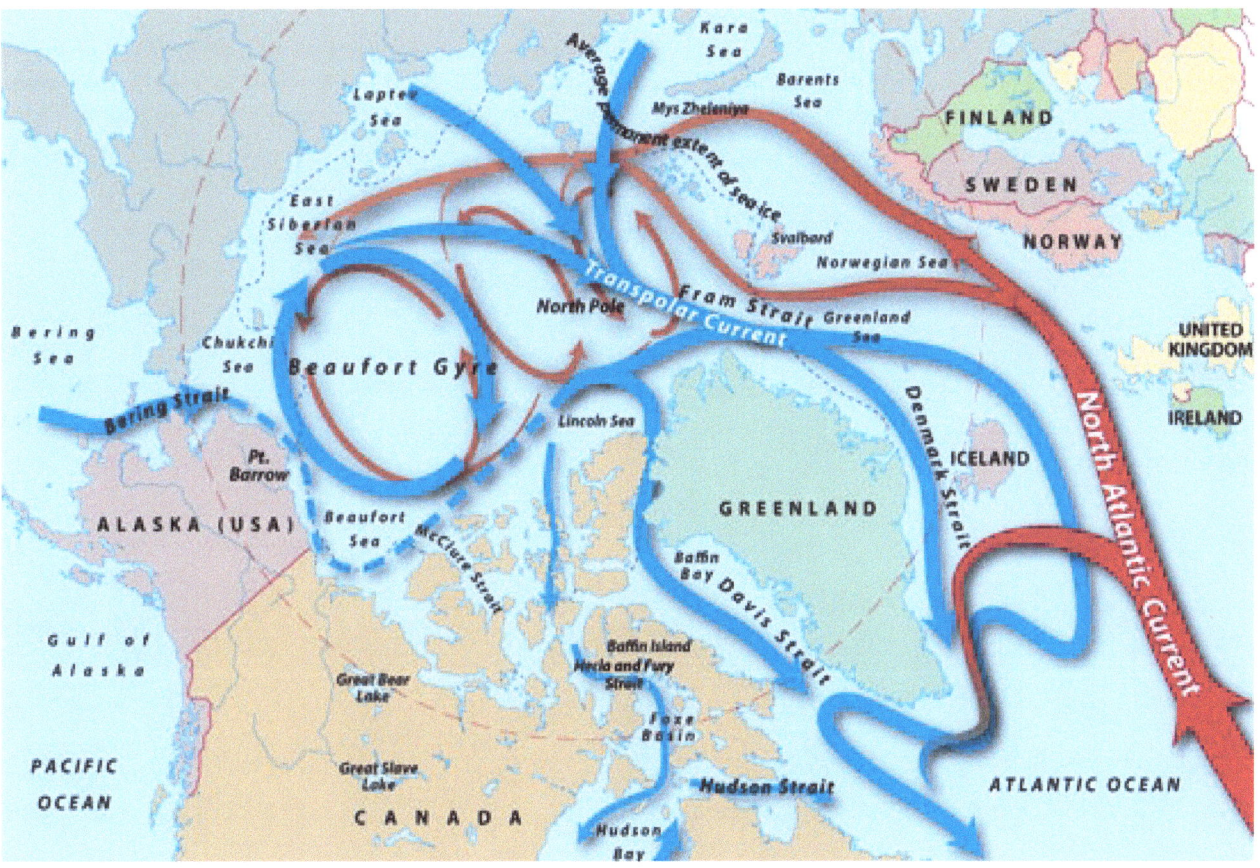

Map Courtesy of the Arctic monitoring and Assessment program. Illustration by Jack Cook, Woods Hole Oceanographic Institute.

Likewise, this map has combined all the elements necessary to understanding how the Arctic Ocean currents interact and their relationship at the present time[4]. (Which, depending on Ocean level in the Bering Strait will change rather dramatically.)

Moving now to that subject, the following charts indicate how they have behaved.

Some 3.3 mya, the world temperature chart showed a marked drop in temperature followed by a new series of gyrations settling in permanently with the complete closing of the gap between the two continents of America taking place by 2.8 mya. We can finally begin to understand what the drifting of the continents had

---

[4] The US Navy was the one that discovered how the North Atlantic current dives under the lighter low salt water south of Iceland. It does so via a series of giant whirlpools with as many as seven operating at the same time.

forced on our Planet, a virtually enclosed North Pole Sea with the Atlantic opening the main current in with only a minor inlet from the Pacific and insignificant exits through the Canadian Arctic Islands down the West side of the Greenland Continent.

*****

6. THE HISTORY OF OCEAN LEVEL CHANGES

Ocean Level Rise and Fall Through Time.

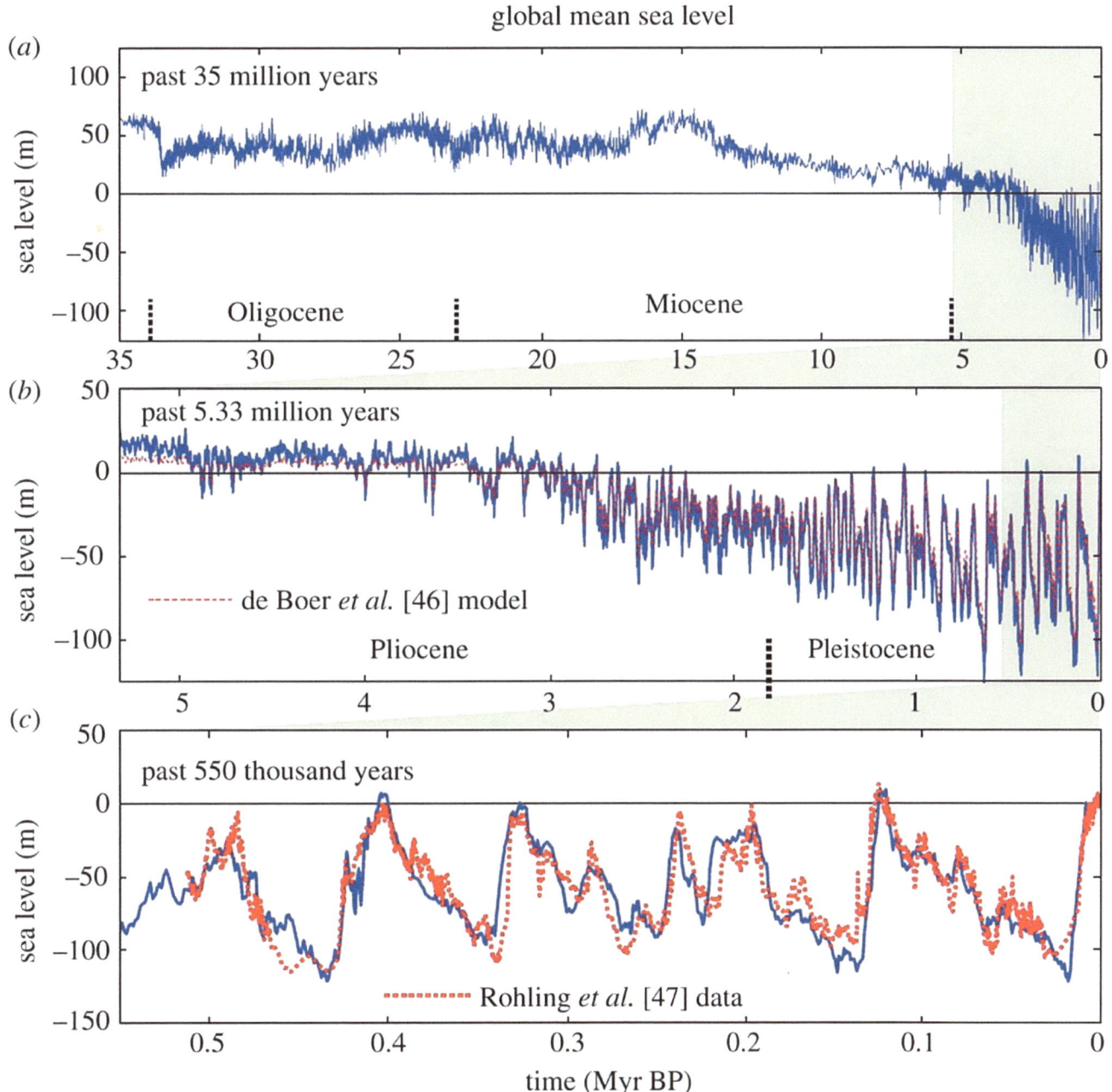

Charts Courtesy of NOAA and Wikipedia.

Not surprisingly they respond to the rise and fall of temperature and in turn whether or not ice is building up on dry land.

Not to be totally ignored is the Continent of Antarctica, which is the repository of the giant's share of fresh water at this time, and therefore the future supplier for much of the world's remaining Ocean level rise potential.

The Home of our Permanent Glaciers-The Antarctica Continent.

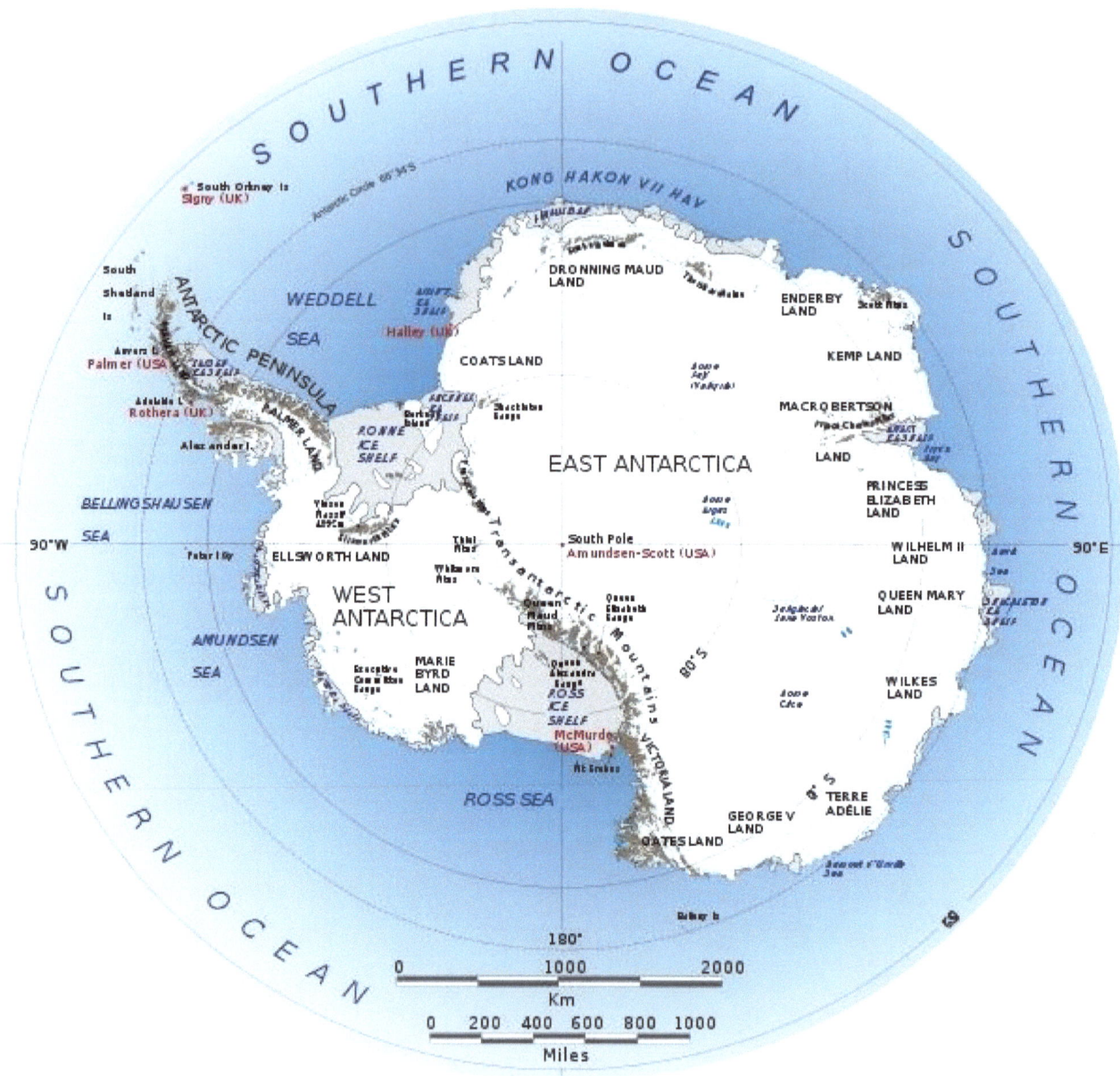

Map Courtesy of NOAA and Wikipedia.

The land portion of Antarctica is larger than that of the Continental USA. In addition, it is surrounded by open Ocean with the ocean currents from the Pacific having no on/off switch as in the Arctic area.

The following image was chosen for its depicture of two vital pieces of information when thinking about that area. First, much of it today is shallow Ocean with the Bering Strait only 100 meters deep along with nearly all the passages between Northern Canada's Arctic Islands similarly shallow. Thus when Northern Glaciers are at their peak at 2 miles deep, the down-thrust on the earth below is so great that it sinks in places by almost a like amount with land up-thrust in the surrounding area almost as great but more widely spread out. Thus in many areas what is today sea, tomorrow becomes dry land.

A Visual Depiction of the Arctic's Land and Ocean Topography.

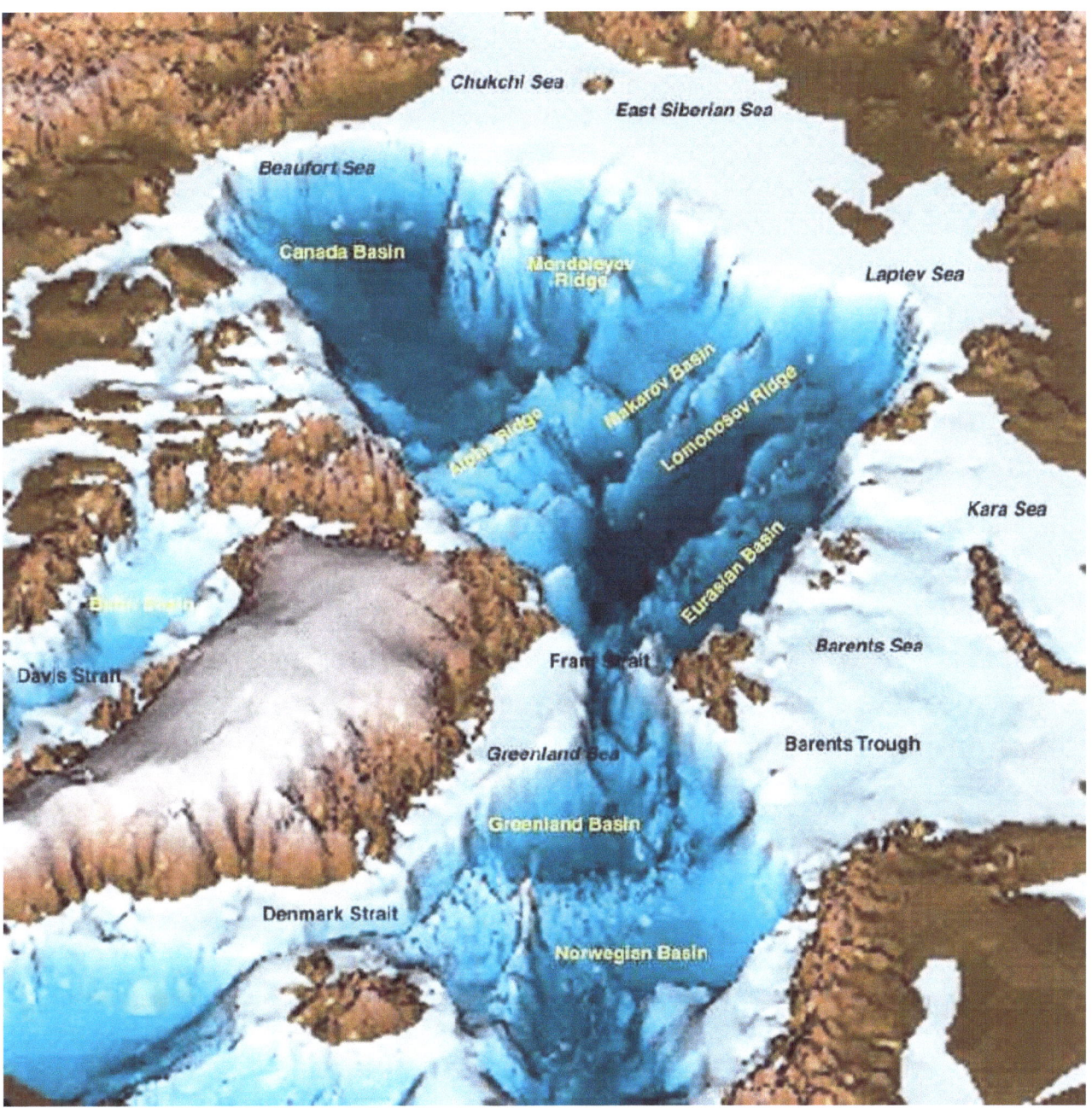

Image courtesy of MASIE Arctic Ice Dataset.

Thus, when we reach the peak of warm periods, the warm water enters this ice-free sea on top from both the Atlantic and Pacific Oceans creating perfect conditions for winter snow…a vast lake-effect that puts that of Lake Superior to shame. To put this into perspective, the Great Lakes cover an area of 95,160 square miles while the Arctic Ocean covers an area 57 times greater or 5,427,000 square miles[5]. Canadian researchers have long known that in the early stages of ice-sheet build-up, the way it forms is that snowfall is so deep that it does not have time to melt before the next winter comes. The consequence is that with a white surface, so much heat is reflected back into space that the snow and ice build-up is actually enforced. (The albedo effect). As the snow and ice gradually builds-up on land, so then does ocean water-level fall, such that, by the time we reach the glacial maximum, that water drop has reached 100 meters below today's level. That in turn restricts the flow of warm water into the Arctic allowing solid ice sheets to form and shutting down lake effect snow and in fact reducing it to almost zero. We then know what happens; in cold winters and the ultra-dry humidity that results, evaporation of moisture from the snow and ice combined later with temperature rise, results in fast ice and snow melt such that the ice sheets start to recede leading to what the Canadian's call 'snirt', or the mixture of snow and dirt which leads to lower heat reflection and faster melting. Indeed, anyone who has lived in the far North (Edmonton), knows how snow and ice rots with warmer air and how the rate of melting accelerates as that process continues until one day it suddenly is gone. The above recitation is not fiction; it is simply a retelling of the facts behind how winter comes and goes for those who are familiar with weather in the far North.

Knowing how the North transitions from snow and ice build-up to one of melting does not tell us what drives that change. However, before getting into that, we have to explore another issue, something that adds one more layer of confusion, and is again work done by our researchers to the North.

Ice Build-Up on Land Versus Ocean-Level Drop

For this part I would like to suggest to my readers that you pretend you have a water-bed, but not one filled with water, but peanut butter. In addition, imagine that your bedroom is in a small swimming pool (filled thankfully with water and not peanut butter) with the peanut butter bed taking up one third of the area. Now the hard part: imagine that you are a large person with your body mass equal to 10% of the volume of water in your room. Climbing slowly from your pool onto your water-bed, you can observe what happens. The water level around your bed drops rapidly but not much else. After a few hours though, two things will begin to happen; first you will notice how you are gradually sinking into your bed with the area of your bed around you starting to rise. Likewise, the water level around your bed will also be rising, but not nearly as much. By the next morning, one will have sunk almost completely with half of one's body below the level of the bed's surface. If one had a measuring tape one would have noticed that the water level around the bed had gradually risen by just over a quarter of the amount your bed

---

[5] Canada-Sweden Polar Expedition.

had risen (A slight offsetting effect due to the land under the water also rising slightly). Now you have the complete visualization of what happens in an 'Ice Age'. First the Earth's 30-75-mile thick crust bends under the weight of ice being added. More slowly, the softer magma under the Earth's crust also gives way, but not evenly, for the thinner parts and cracks in the Earth's crust start to fill from the upward pressure of the magma under the crust.

The other aspect of this rise and fall in the Earth's surface is the timing, which of course is greatest during the icefield's weight change. But as we can see from this chart, that change is still going on some 15,000 years later As it is, in the North Bay area of the Great Lakes, in the 10,000 years since the ice fields disappeared, that chart shows a rebound of 500 feet with the earlier stage undetermined but showing an almost vertical climb.

The Great Lakes Area Land Rebound Data.

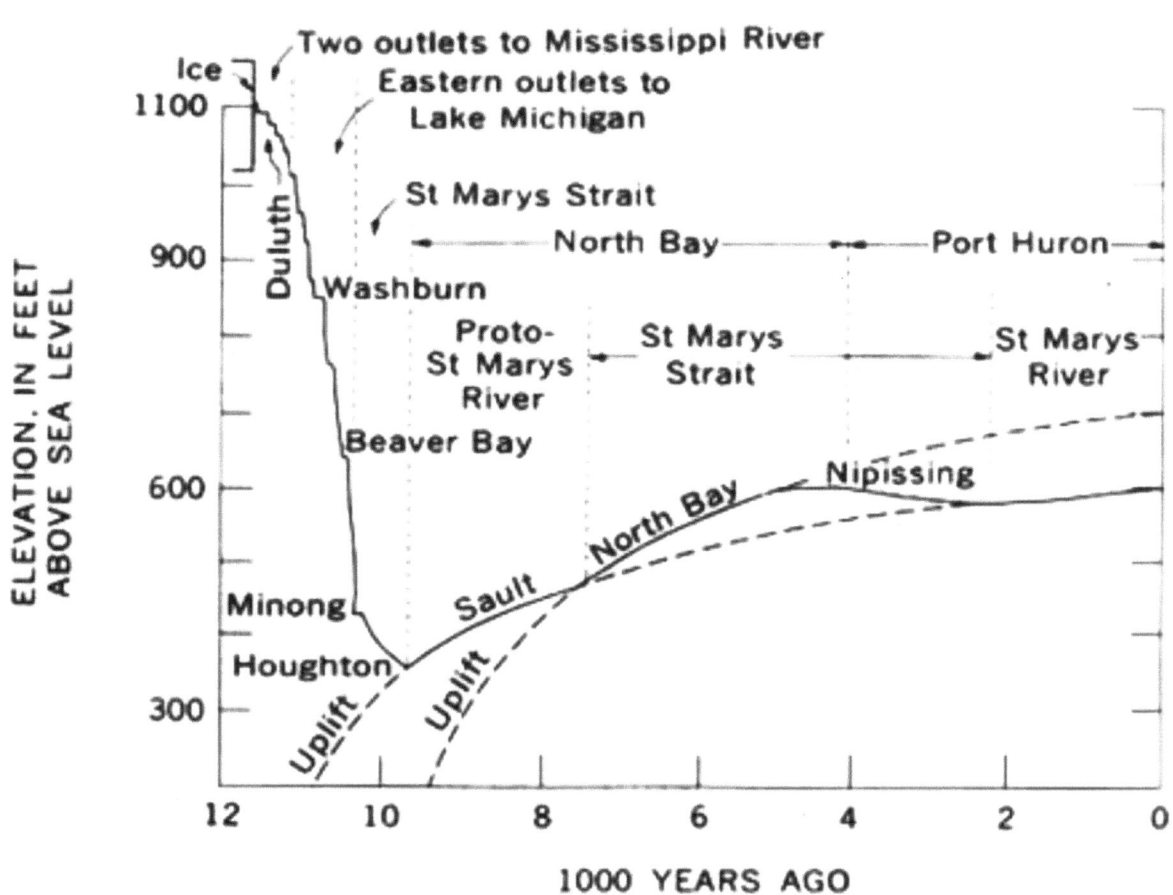

Chart Courtesy of Wikipedia.

If anyone is under the illusion that the above might be an exceptional example of land rebound, a recent article in the New York Times describes a 3 inch per year rebound rate still going on in Juneau, Alaska! For anyone interested that translates to 250 feet per thousand years and 2,500 feet in ten thousand years, and let's leave it at that, for it's almost beyond one's imagination to picture such a change. For myself, the definitive description of what we are trying to explain is what happened in Northern Canada (The Northern Great Lakes Area) where a two-mile deep (3km) glacier formed and was called the 'Laurentide Ice Sheet.' Almost inconceivable was the fact it alone lowered the World's Ocean levels by 230 feet (80m) versus the total drop of 400 feet (120m). That weight shift of seawater translated into adding 3520 feet (1072m) of rock onto that vast area, and if equilibrated over time meant that whole area would drop by that much with the surrounding area being forced upwards by an equal amount. Contrary to the models being used to estimate that rise, I would suggest that they contact Bristol University in the UK to develop a more sophisticated model that reflects the upward pressure so that we can determine where and when it would have taken place, for of one thing we can be certain, it would not have been equal, with some areas rising both faster and higher than others. That, in turn, would translate into a better understanding of how both earthquake and volcanic activity might have been affected by this massive transfer of weight.

As to the other areas that were also experiencing huge weight transfers, the 'Fennoscandian Ice Sheet' in Northern Europe also caused the land under the North Sea to rise forming a land bridge between the UK and mainland Europe. And, of course, Antarctica, with its 91 volcanoes would have made for an interesting time in that area. Also, the area affected by land rise was the sub-sea land between Antarctica and S. America where all, but the Drake passage was blocked-off reducing the Ocean flow from the Pacific to the Atlantic Ocean. Interestingly, one wonders if the huge volcanic explosion of Mount Dawson in 2000 BC was caused by that land drop.

I left the best to the last. The Bering Strait, which also was squeezed upwards and forming another land bridge between N. America and Asia, but more importantly stopping the flow of warm sea water from the Pacific Ocean into the Arctic Ocean. No more would 'Lake Effect' play a part in winter precipitation falling in Northern America and Eurasia. We now know the mechanism by which the Ice Ages both begin and end. It is the combination of Ocean rise and the warm currents entering the Arctic Sea followed by their cessation as Ocean currents are cut off by land rise stopping high winter precipitation from the 'Lake Effect' taking place. Unknown at this time is the role Antarctica was playing in terms of the temperature and strength of the current flowing northwards, for it too could have had a major influence on what was occurring.

Now, I think we are ready to try and understand what was driving these changes which happened long before our modern civilization appeared.

Simple Answers to Simple Questions.

We now come to a series of what we might call 'common-sense questions,' for true scientists at this point will wait for definitive readings to give them the answers with the proof at hand, despite the fact it will be too late for any of us to take action. The questions are few and straight forward.

1.) What triggers the start of an 'Ice Age'? The rise of Ocean temperatures and Sea levels such that Ocean temperature-balancing currents slow to a halt.

2.) What determines the temperature level at the bottom of that cycle? The level of CO_2 in our atmosphere.

3.) What triggers the start of a 'Warming Age'? The fall of Sea levels such that the Arctic Ocean is insulated from intermingling with other Oceans and freezes over.

4.) What governs the length of these cycles? The up to 50,000 years it takes for the rebound effect to take place in each direction or 100,000-year cycles.

5.) Can that cycle be shortened? Yes, by increasing CO_2 levels in our atmosphere with that number recorded in our history. (Somewhere between 400 and 800 ppm).

6.) Let's say all the above answers turn out to be correct, does this solve our problem? The answer is a resounding no, for although I expect this hypothesis might be correct, we need the definitive answers on such a serious issue. Luckily that answer is just around the corner and so please read on!

*****

7. WATER

What better way to introduce a subject than by way of describing the two kinds on the Planet? We all know the basic inputs for most life on Earth is oxygen, carbon and water, which when combined forms most of the products upon which life relies. Sneaking into this mix comes Sodium Chloride, washed in by the rain as it fell on dryland with that product being highly soluble in water. It was only when life later crawled onto dry land and adapted itself to rely on unsalted rainwater that a problem presented itself. There wasn't much of it. So, how do we show that fact? Luckily the USGS has answered that Question;

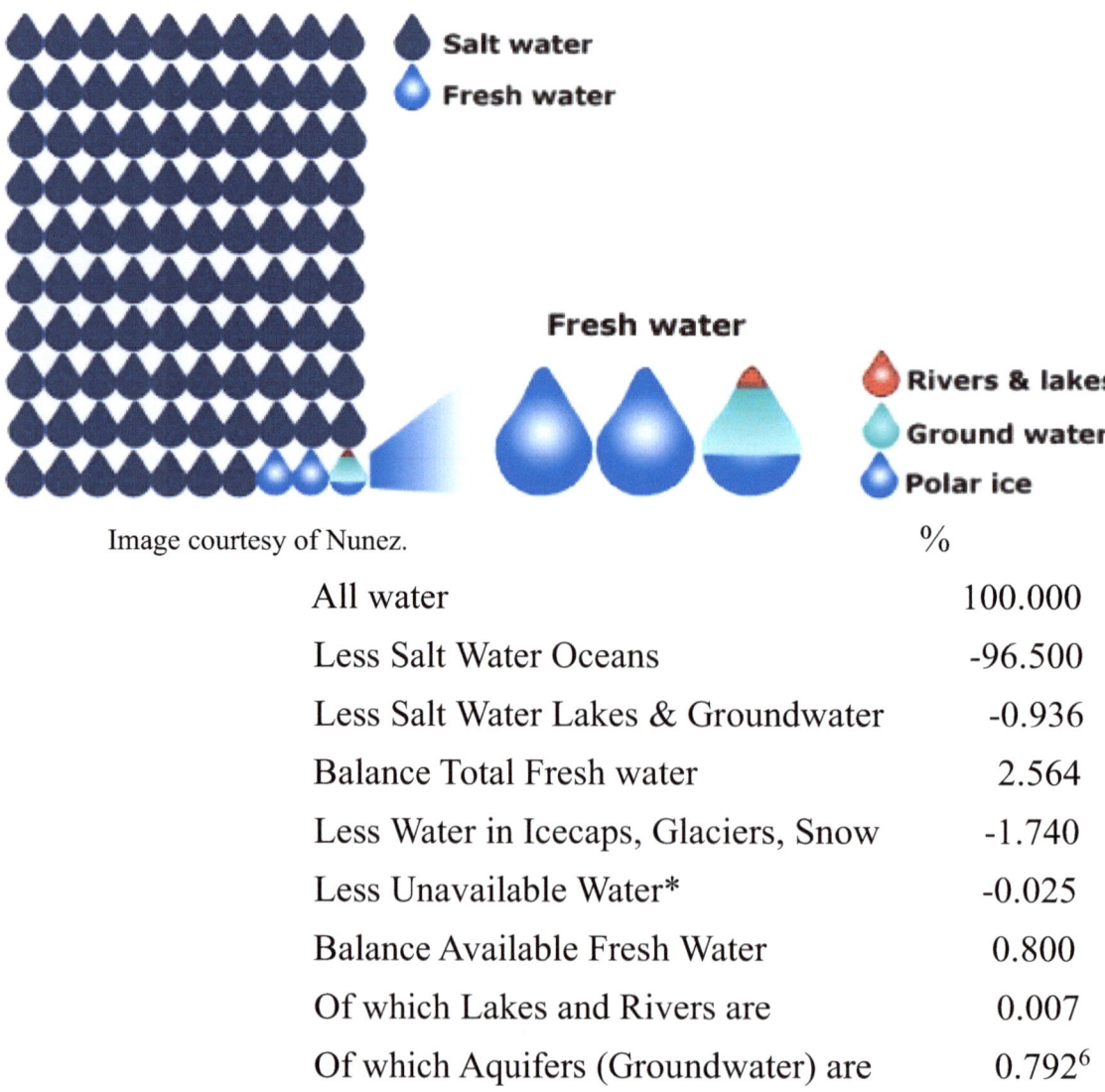

Image courtesy of Nunez.

	%
All water	100.000
Less Salt Water Oceans	-96.500
Less Salt Water Lakes & Groundwater	-0.936
Balance Total Fresh water	2.564
Less Water in Icecaps, Glaciers, Snow	-1.740
Less Unavailable Water*	-0.025
Balance Available Fresh Water	0.800
Of which Lakes and Rivers are	0.007
Of which Aquifers (Groundwater) are	0.792[6]

---

[6] Reflects recent discoveries plus in transit movement of underground water.

• Soil moisture 0.001%, Plants and animals 0.0001%, Swamps and marshes 0.0008%, Soil ground ice and permafrost 0.022%, Atmosphere 0.001%.

Glaciers and How They Melt

The World has four massive and long mountain ranges; The Rocky Mountains, the Andes, the Himalayas, and lastly there is one last huge, but diverse range that stretches from the Carpathian Range in Central Europe to the Aldanskoje Nagorje in Eastern Russia. It too is a vital source of water for the many countries that surround it and share this common story.

All four contain, or contained, thousands of ice fields called Glaciers, which were built up from snowfall in winter and are returning it as melt water into rivers during the summer ice-melting season. This crucial source of river water is about to change and in a rather dramatic way.

While it may sound obvious, it is key to recognize that Glaciers are thickest at their bottom and shallowest at their top. As a consequence, how many feet a glacier shrinks in a year only tells us if it was at the bottom or top of the glacier.

As an example, from 1963 to 1978, the Andes glaciers were shrinking at the rate of thirteen feet per year. By 2016, that number had changed to two-hundred feet. All that means is that those glaciers are nearing the end of their lives. In contrast, the Athabasca Glacier in Jasper Park, Alberta, Canada only retreated a few feet this year, but dropped nearly ten feet in height versus the eighty feet since 1958. In other words we can expect a more rapid increase in it's rate of shrinkage simply because it is getting shallower in depth.

Due to a combination of shallowing and rising temperatures, today, we have higher ice melt rates than normal. Thus, when these glaciers are gone, so will the river flow decrease be even more dramatic. In all four Ranges, the story is the same with forecasts of the smaller glaciers gone within five years, and the bigger ones within twenty-five years. Wait! What is about to happen is worse by far than that, for all the Ranges share another three things in common. The first is that without that source of water in the summer, it coincides with it being the low rainfall season and the most sensitive time for crops to be without water. At this same time, we also have to worry about the needs of human population centers. Lastly, in all these areas, many electrical generators will also have to shut down. Meanwhile, different countries are arguing about diverting rivers, building dams and diverting water use from farming to cities and towns in a mad scramble for what may be about to disappear. Given the fact that the supplying rivers will probably be drying up by then, it does seem like a pointless debate, never mind lost time in an effort to build up rain storage facilities.

The first question this chart raised was had it changing with time, and if so, by how much? Well, it turns out, the answer was yes and by a lot. It was all about coming out of the last ice age and where we are in that cycle. Luckily, we seem to be in a warming part of that cycle with humanity giving it an unintended shove in that direction. So, what does that mean?

With most of our rivers dependent on the summer meltwaters from those glaciers, knowing how long they are going to last is important to say the least! Think about it for a moment; glaciers are the last remnant of fresh water left from the last ice age. Not only that, they are the last recharging source for many of our aquifers and for most of our rivers, providing them with their summer flow of water. With the variability of winter snowfall, one can only wonder what lies ahead for many areas when future summers come.

Just what did all this mean in terms of what has happened since the last 'Ice Age'? My private guess is that if we added together the three great 'Ice Sheets' that covered Canada, Scandinavia, and Russia, they would have doubled the number in the above chart to perhaps 3%. Put another way, an ice age is a huge ice-making machine that stores fresh water in the ice form for the future in a time when the Earth warms up and that is where we find ourselves today. The message is we need to conserve these sources for fresh water may well be our 'Achilles Heel'.

One last thought, there is a huge difference between an 'Ice Field' and a 'Glacier'. The 'Ice Field' is more like a pancake that has been poured over a bumpy surface with the thin parts being on the mountains and the thick parts being in the 'Glacier' valleys. Hence the 'Glacier' probably contains most of the frozen water, but not always and by no means will it melt at the same rate as the 'Glacier'. Using the analogy of the pancake, the thin parts will melt first and more quickly with the solid ice parts, the 'Glacier', more slowly, for it also may be shaded from the heat of the sun. As an example, in Canada's 'Columbia Ice Field' the area covered has shrunk by one third in 20 years, but that does not mean that the 'Glaciers' have also melted at that rate. In fact, without having access to the ice volume loss data, I suspect it would be 25-50% less, depending on the amount of meltwater in contact with the 'Glacier'. An excellent example of this phenomenon can be found when looking at the 'Ice Field' that supports the 'Gangotri Glacier' in the Himalaya Mountains, and is the prime source of water for the Ganges River of India. There, common sense tells us that the vast 'Ice Field' will melt first, but the main 'Gangotri Glacier' might last twice as long and well into the next Century. The downside of that is that the flow rate of ice melt into the River itself would probably by then be a quarter or less of present flow rates, if behaving similarly to the one in Canada.

The River and Lake Fresh Water Supply.

First, some history; with the exception of the occasional natural spring and assuming no containers, early humans would have been forced to settle close to rivers and lakes. One can now see the problem that humanity faced from earliest times, for the visible water was only 0.0072% of the total. It sounds simplistic to say that all early humanity lived within one day's travel from fresh water, but that was largely true with spring water the only exception. And now to some history which might explain how we got into this fight over water and its uses.

Early man would have noticed that without water most plants would wilt and eventually die. Hand watering was the only thing that kept them alive and so irrigation from small streams with rudimentary dams and canals was the answer for early farmers. I would have liked to have found the inventor of the first dam to control water flow, but almost certainly that honor goes to that lowly rodent, the Beaver. Common throughout Europe and Asia, farmers and hunters there would have been awed by their skill in building dams and using the ponds they formed to protect their homes which they then built at the center of those ponds. In any event, the first recorded use of levers (the shadoof) to raise water was in Egypt and Mesopotamia with their being accompanied by the use of simple canals to conduct water short distances. History records that use as being sometime after 5,000 years ago.

Egyptian Tomb Image of an early water lifting device (shadoof).

Courtesy Water Encyclopedia of image from circa 2000 BC Egyptian Tomb.

In 2002 BC, Joseph, of biblical and Koran fame, was the first to build a huge 150-mile long gravity-fed irrigation system[7] that today still supplies water to the Nile Valley's farmers. Included in this masterpiece was a system of dykes that later became roadways from the Nile River to the western wall of the Nile Valley. These allowed farmers to flood their fields with time for the fresh silt to settle on

---

[7] Joseph, Moses, and the Exodus, 1917, by Fred Graham-Yooll.

their fields before the dry summer set in.

Later, when a subsequent Pharaoh had planned to run water into and out of the nearby Nile to fill a vast depression called the Faiyum, and to convert it into a vast reservoir that would act to reduce flooding, they discovered, to their horror, that the two ten-mile stone-lined canals they had built were far higher than the water level of the Nile. Concluding it was not feasible to make it deeper, his advisers turned to the now 90-year old Joseph for a solution. His answer was to extend his irrigation canal to the Faiyum such that it became the way to turn that natural basin into a lake, and in dry seasons, raising back the level of the Nile so that farmers in the Nile Delta could again irrigate their land. Luckily, I discovered the hard evidence that proved what Joseph had done and its timing[8]. As an aside, the Lake became a large source of fish such that the taxes from that activity paid for both the upkeep of that system and Joseph's earlier irrigation canal, itself an incredible achievement, given the rudimentary knowledge of hydrology. For the next 4016 years, those farmers were able to supply water to their crops ensuring some of the highest agricultural productivity in the world. Only in modern times have the people of Egypt interfered with that system with declining yields the result. Incredibly, even though this is the World's oldest working system of any type, it has been neither recognized as a World Heritage site, nor given the recognition that it is over 4000 years old and still in existence, never mind still in working order! Another, but unprovable innovation by Joseph was the widespread use of hand dug wells in Egypt's Delta shortly after his move there.

The Persian Qanat irrigation system came next. Building a series of tunnels that captured the melting snow from the mountains of what is today NW Iran, they carried water to the many fertile valleys and desert areas that were to be found to the south and east of that source. Other than surface salt build-up from faulty irrigation practices[9], that system is still working today.

Next, shortly after 400 BC, came the famous Egyptian Aqueduct in Alexandria to transport water from the mainland to the island of Pharos. Along with its famous systems of canals, underground water supply and sewage systems, not to mention showers, water clocks and other devices, and finally, with its teaching Library, it attracted no less than Archimedes to make it his home. "And now you know the rest of the story!"

While Rome made the Aqueduct famous for bringing fresh water to Rome and its cities around their Empire, the credit for inventing that technology belongs to the Minoans some 1,000 years earlier.

Despite other developments begun in China, the Philippines and in the Andes by the Aztecs, such as sunward facing stepped terrace irrigation systems carved out

---

[8] Joseph Moses and the Exodus by Fred Graham-Yooll
[9] 1953 Survey of Persia's Qanat system in which I participated.

of mountainsides, with each having walls so that water could be retained in those terraces to form ponds in which rice and fish could be raised, the ultimate prize must go to the Khmer.

Deepwater or Floating Rice

Courtesy of Huynh Ngoc Duc.

In AD 800, in what is today Cambodia, Viet Nam, and Thailand, the Khmer rulers took a River (the Mekong) that was subject to extensive monsoon flooding, and by adding a series of huge deep and high walled storage ponds for future irrigation use, converted that area into farmland with an irrigation system that not only allowed crops to be grown year-round (deepwater rice, then vegetable crops–all leaving high organic material behind), but also created vast fisheries that provided its population with a new and nutritious source of protein. And now we come to a dilemma all too common in this day and age. Many of today's high yielding crop varieties achieve that gain by reducing stalk height thus allowing more energy to go into the production of the flower and the plant's seed, which of course is what we harvest. The problem though is that this means less plant material is returned to the soil which in turn means lower organic matter and less water and nutrient holding soil capacity–all long-term negatives in the grain growing areas of the World. If we thought that was bad, think what replacing deep-water or floating rice as it is called, with dryland high yielding rice does. Lost is not only all that stalk organic matter, but also the material from growing vegetables and the fish life that depend on water and supply protein to the people. Also, being lost is a system of long dykes that selectively both stopped flooding and guided flood-waters to fill those storage ponds. A wonder of its time, it has been named by UNESCO as a world biological heritage site. As an aside, storage ponds were built around large buildings such as Angkor Wat, giving the soil under them the hydrological support to stabilize the ground under them. With the storage ponds largely abandoned, the old ruins are now starting to seasonally heave and fall, splitting them apart (see below).

Destruction of Angkor Wat Ruins from ground heaving.

Courtesy of Wikipedia.

We are not done yet with what has been happening in this area and also in Egypt and many areas of our World. The discovery of electricity, and the fact it can be generated by water being stored at higher levels behind dams, has created a new demand for water. While some claim that this new demand is compatible with that for irrigation, and the two being able to co-exist in harmony, that is more of a political claim than a fact of life, for agriculture needs storage of water to supply irrigation water in the dry summer season, while electricity needs storage of water to level out the year-round flow making that source reliable for use by industries and cities. The conflict is obvious never mind the other potential negative impacts there might be to wildlife or fish. On the political front there is a commonality in all these decisions. With city dwellers the visible voters and the wealthy, electricity is a must for them and their style of life, and so the decision is usually to build dams for electricity. Then when people see the damage being done to agriculture, fisheries, and the environment, the hand-wringing begins. Too late is the usual cry, but then comes education with the electorate demanding action which may or may not follow. The World though has taken a giant step backwards in terms of it being able to support humanity. This despite the fact we have in one-hundred-and-twenty-five-years gone from one to eight billion people on our planet, a growth rate that even optimists must shake their heads at in fear. Even worse is the fact that even if family size levels out at two or flat worldwide, because of the young age of our population, growth will continue to the eleven range by the end of this century, an unsustainable number spelling destruction of Earth, never mind stretching its resources past the breaking point.

Moving now to what 'Global Warming' is doing. The obvious point is that most of the world's key rivers start in our mountains and are fed from their Glaciers. With our experts advising those Glaciers will largely be gone within ten to twenty-five years, one wonders what our World's leaders are thinking? It spells out that water flows will be dropping by as much as 50% in the big rivers that supply our major population centers throughout the World. There is another catastrophe hiding in that figure. Our rivers are already being treated as open sewers and dumping grounds and if that continues unabated, the pollution levels, just from the arithmetic alone will double! The appalling fact is that many countries today are allowing raw sewage to be dumped into their rivers with Venezuela and Brazil total disasters followed by even Washington DC when it rains. As for Asia and Africa, many countries there have not even made the slightest effort to take care of that problem.

So, our world forecast for glacial sourced rivers is for a reduced fresh water flow of 50% with an increase in pollution levels of over 75% (50% from the mathematics and 25% from population growth) in less than 25 years. For the few non-glacial sourced rivers, our forecast is for a twenty-five% reduction in water flow and a 37.5% increase in pollution.

What this means for the creatures living in our seas is almost beyond comprehension.

The Underground Fresh Water Supply.

It helps to understand geology in order to appreciate how the world's aquifers were formed, for the pressures and temperatures at depth are extraordinary. While gravity allows water to penetrate downwards into the cracks and crevices one finds in soils and rocks, they in time, as they move deeper under the surface, find those pockets of gas disappearing as the pressure becomes so great they actually start dissolving into the rocks themselves as they are converted into magma. And that is why underground rivers are rare and almost always shallow. In turn, the cracks and crevices one finds there are caused by ground movement as our Continents drift around our World and our ice ages come and go. Thus, they tend to be short in length rather than offering long spaces down which underground rivers can travel. On the other hand, sand and limestone rocks can allow seepage of water to take place with that very action often dissolving or eroding the rock and thus creating channels down which the water can flow more quickly. Conversely, when water encounters hard volcanic (igneous) rocks, which are non-porous, those fields tend to be short lived and when empty, never refill as is the case with oil or gas deposits. In all cases ultra-deep water needs to be handled with care for not only is it under enormous pressure but, is hot and usually contains some unpleasant products.

The next thing to appreciate is that the location of most of these aquifers was dictated by a combination of where the porous rock was laid down and what was its shape or angle of slope. As mentioned, in the Northern hemisphere over the past three and a half million years of 'Ice Ages', with some of those Glaciers being over two-miles thick, that caused a depression in the earth's crust to form of hundreds, if not thousands, of feet in depth, with rebounding taking place as that ice melted. Conversely, the land is lowering to the South as that pressure from the weight of ice disappears. That process took tens of thousands of years to complete with the rebound in Canada and Russia still in transition. Thus, in Canada, all its rivers flow North to the Arctic or Atlantic with only the Rivers to the West of the Rocky Mountains flowing into the Pacific Ocean. Conversely, all USA rivers flow South except for the one in the Red River Valley in N. Dakota, which flows into the Hudson's Bay. Thus, the dominant slope in the USA is from the borders of Canada to the South with that underground water flowing and seeping from that direction at relatively shallow depth and doing so all the way to Texas and New Mexico in the South. In turn, most of that water came from ice age melt in the North which may be gone within the next twenty-five years with the melting of the 'Columbia Ice Field'. Coupling this with the seepage rate being anywhere from feet per minute to inches per century, and one gets an idea of there being no such thing as an average aquifer.

Another surprise to many is just how recent is the use of groundwater. This despite knowing about the presence of naturally flowing wells and springs for centuries. Indeed, the first wind-pump was not even invented until Daniel Halladay did so in 1854. With a lifting capacity of a few gallons per minute, it was ideal for supplying water to livestock and homes. Thus, even small sources of underground water were adequate for that use with over 600,000[10] in operation by 1930. And so began the use of underground water, be it in only small quantities.

It was not until the 1890's that the USGS (N.H.Darton) discovered the first Aquifer near Ogallala, Nebraska, and with later research discovered it covered the USA as far as Texas and half way across the USA Plains. This huge fresh water asset remained undeveloped despite the 'Dust Bowl' (1930-36) going on above it,

---

[10] Halladay records.

due to the lack of technology to move it to where it was so badly needed. After WWII, and with so many surplus gasoline engines left over from that war and available at discount prices, did that become the popular way to lift water from the Aquifer.

A Modern Center-Pivot irrigation system.

Courtesy of ATS Irrigation Inc.

The Center-Pivot from the Air.

Courtesy of Alibaba.

With large quantities of water now available, that event was soon followed by the spread of flood irrigation by farmers. Then, from the 50's to the 70's the development of commercial sprinkler systems heralded in the introduction of the rotating circle type now seen all over the West. Bear in mind it was at this point that our World's population had climbed from 1 billion to 3 billion and was getting ready to more than double over the next 30 years.

Ever the inventive creature, but with no sense that any planning or co-ordination between competing uses was needed, humanity worldwide turned to underground water for its needs. Seemingly with endless supplies, those demands kept increasing with the World's population passing 6 billion in 2000, and only then were voices being heard that the water level in many of these aquifers was dropping and the end was in sight for their use. So too were water supplies being stored in Glaciers, Ice Sheets, and Snow-Packs around the World as 'Global Warming' was taking its toll with the impact of this development both hurting 'River Flows' and the replenishment of our 'Aquifers'. Finally, our experts started to investigate the question that should have been asked at the very beginning: what was the source of all this water and whether it was a 'onetime shot' or in fact could supply the 8 billion people now here on Earth? It was now 2018. We, of course, know it came from the melting of all that ice from the last 'Ice Age', but the question of its renewability was much more complicated with it covering a whole range of possible sources. But again, they all led back to that source. With N. America having used this source the longest, not surprisingly, with its large agricultural base and heavy use of aquifers for irrigation, most aquifers are in decline with the Ogallala already down to only 50% remaining.

World Aquifers and their stress levels.

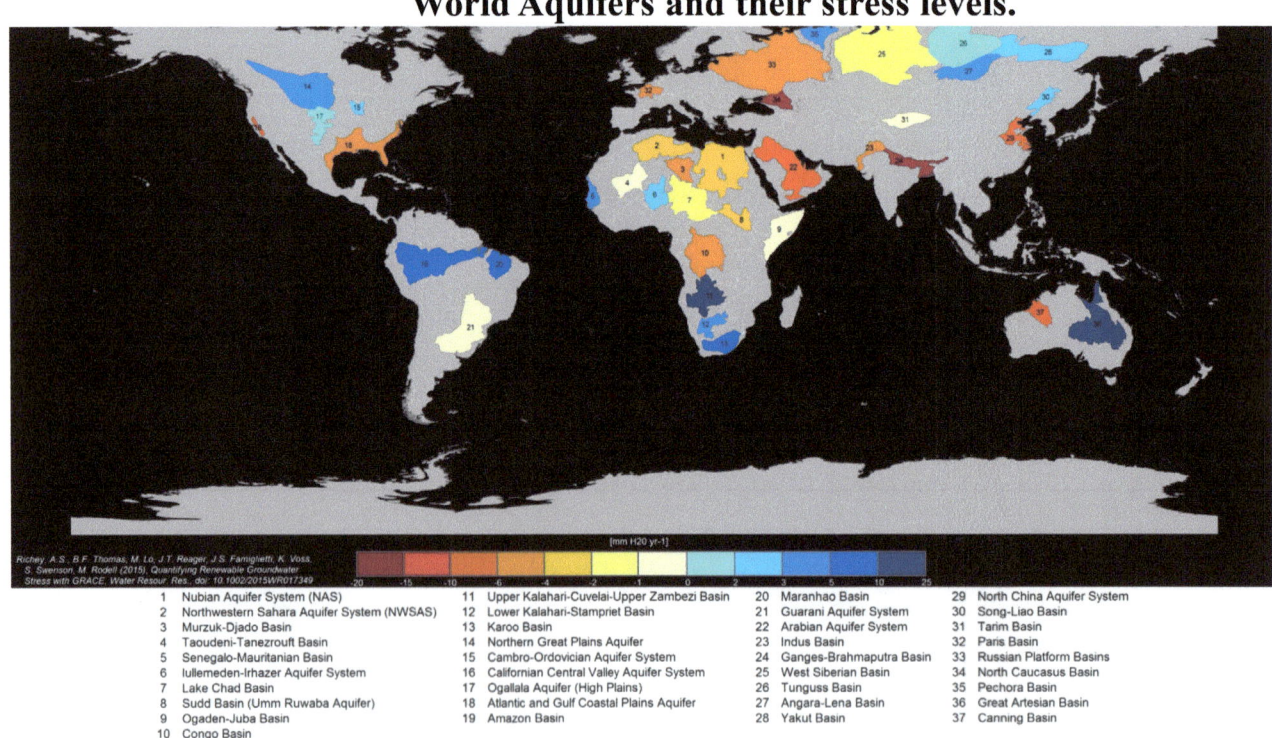

Courtesy NASA June 16, 2015

In all of this, one message comes out loud and clear, agriculture and industry are the villains with their use of water gobbling up nearly 90% of the amount being withdrawn from aquifers. The question now is what comes next? Luckily the answers around the World are the same for everywhere.

The Principles

There is a beauty to simplicity and if any industry can exhibit that trait it might well be Agriculture and Horticulture. Farmers have a habit of going straight to the heart of the matter which is; if it doesn't make sense, don't do it! Well, trying to raise more grain, corn and soybeans by smothering it in non-renewable water is crazy so either stop doing it or change to some other system that matches what is being replaced! The steps are the same for everywhere:

1.) Replenishment of organic matter in the soil (7%) so it can retain more moisture for crops to grow (needs less frequent irrigation). May require the insertion of a grass legume mix into the crop rotation in order to increase soil organic matter.
2.) Increasing the surface trash cover to reduce water evaporation from the soil (reduces frequency of irrigation).
3.) Install the latest software and measurement devices to optimize water use and crop uptake. (Again, reduces both quantity and frequency of irrigation).
4.) Grow only drought resistant crop varieties (Same as above).
5.) Construction of regional rain retention ponds to replace aquifer withdrawals.
6.) If that fails, a switch to drought resistant crops or fallow in the crop cycle (Requiring no irrigation).
7.) If that too fails, conversion of the land to pasture, reforestation or solar cell energy production should be explored.

A mandate that imported products are banned else none of the above will be done! (In order to make sure food supply is secure and economically allows domestic producers to make the above changes). That simple mandate, rather than a series of almost unmanageable regulations, is the practical way to ensure water efficiency by farmers.

For Horticulturalists, the list is somewhat different for the growing of specialty crops, fruits and vegetables requires a different set of guidelines. Ranging from the use of greenhouses, drip irrigation, and hydroponics (to reduce water use and loss to minimum levels) these systems are some of the most efficient in the world. Likewise, they also require a mandate on domestic production, for other developing countries do not require these kinds of efficiencies with their associated investments and costs.

For Fisheries, the opportunities are very much tied to the storage of rainwater in a system of lagoons throughout the country combined with hydroponics and other water systems (water conservation and flood control).

For Forestry, the challenge is to change the types of trees being grown as climate changes take place. Indeed, with land often under Government, or no control and ignored, its care is often a 'hit, or miss,' situation with 'slash and burn' making a mockery of that profession.

Fresh Water Demand

World demand for fresh water is remarkably consistent with the latest United Nations estimates indicating 70% being used by agriculture for irrigation, 23% by Industry, and 8% by municipalities. These are gross numbers and so we can address this demand by applying some generalizations.

-The most advanced Agricultural and Horticultural irrigation users are already using software that measure soil moisture and crop conditions in order to minimize the use of water. These programs not only increase crop yields but do so using up to 25% less water than previously. Experimentation is also being done on methods to reduce water loss by evaporation. As for hydroponics and the genetic engineering of new varieties of plants, the sky is almost beyond what we can imagine. As a long-term soils and fertility expert, I for one am still amazed at what the future might hold. On the negative side, our politicians around the world pander to their dominant city electorates giving little attention to agriculture and are continuing to make decisions adverse to that industry. The surprise is that there appears to be little difference in that decision-making process between some of our most advanced and backward civilizations. As an example, California in the USA is making decisions on water use and supply that are both short sighted and absurd while ignoring the simplest steps of optimizing supply. On the other side of the world in SE Asia, a highly efficient agricultural system is being destroyed, all to make room for hydroelectric production! In between this are all kinds of other examples of both good and bad decision making with the prime example of the good being New Zealand. For a balanced and ecologically sound agricultural industry, New Zealand would be hard to beat. (Their balance of crop and animal production is still a joy to behold.)

The bottom line of all of this is that the good things are largely being offset by the bad, and with less groundwater being available in the future, agricultural production will at best remain relatively flat for the short term. Only when hunger and starvation threaten will our politicians start making the decisions that will increase agricultural production with gains of 25% possible. So, how are the fresh water sources of supply doing? The fact that good water statistics date back to 2010 is an indication that the world is still not taking this problem seriously.

The Decision-Making Predicament

All that remains to be discussed is what Governments do or don't do to our environment, and that unfortunately usually means one of two things: either not doing something they should have been doing, or something they shouldn't have been doing. Luckily, I only need a short sentence to explain that quandary. You don't have any capital[11] (money) to spend! If you want some you have to sell something to someone to get it; either that or you have to print it, and if that too is impossible, borrow it, that is, if you can find a lender! Unfortunately that leads to deferring all but essential investments not involved in feeding, clothing and supplying drinking water to people. Sometimes some weird things can be mixed up in those decisions, such as being forced to exclude water for agriculture to satisfy

---

[11] You need to read my book titled; 'Bumbling, Fumbling, Stumbling in the Dark.'

thirst now, but at the cost of starvation later! Even more bizarre is when water is denied from that use in order to produce electricity for quality of life improvement at the risk also of starvation later. By comparison, and almost inconsequential, is the abandonment of things that jeopardize life itself such as investments in sewage, pollution control, and our environment in general.

The Situation in North America.

It may surprise some to learn that the Banff and Jasper National Parks of Canada are home to N. America's largest 'Ice Field'. Called the Columbia, and feeding eight different Glaciers, the melting waters from them supply all three oceans that surround Canada. The Columbia and Fraser Rivers into the Pacific, the Peace, Athabasca, and McKenzie Rivers into the Arctic, and the North and South Saskatchewan Rivers into the Hudson Bay and then South into the Atlantic. Interestingly, there is only one small river that flows from the USA into Canada, the Red River from North Dakota and Minnesota.

In the East, the Great Lakes is the mother and father of the fresh water in North America containing 20% of the World's lake water. While the overflow from these lakes exits into the Atlantic via the Gulf of St. Lawrence River, that too might change with the land rebound still going on in that area. A long-term drought condition that lowered lake levels dating back to the 1960's is officially over with record rainfalls in 2013-15 bringing levels back up to normal. Despite this recovery, no inclination by either Canada or the USA has been expressed to the idea of exporting water South to the USA. While this might be practical, as long as flooding of the River systems to the South continues, the argument for importing more water makes little sense. Far better would be the idea of constructing large lagoons to store surface water or for the replenishment of aquifers for later use. Not only would that be efficient and ecologically sound, it would also reduce the massive costs of flooding from year to year and be under their sole control.

–Rainfall, compared to many other areas of the World is relatively stable with the biggest variation coming from the yearly status of 'El Nino'.

– Snowpack run-off in summer mostly impacts the Westcoast Area and the northern Canadian area. So far, that has remained relatively steady through the early summer period except in the SW USA, where it has almost reached zero due to the disappearance of their glaciers.

–Glacier melt in late summer is a thing of the past in that Continent except in the NW where it is also forecast to reach zero by 2030. Only in Alaska is that melt forecast to continue into the second half of this century with the annual run-off just in the Southern area equal to 1.5 times the flow of the Mississippi River.

Earth, a Planet in Peril

Forest Fires, Both Natural and Self-Set.

Courtesy of 'Forest Fire Protection Services and Xorex Service who designs the fire detection software.

– Today, with water in the summer coming mostly from glacier melt, the worry is about what will happen to supply when the glaciers are gone. Cities are growing rapidly with their demands for water taking priority over farm use. Short term projects to divert rivers through tunnels will not help as that source will also lose its source of future supply. As for irrigation water, the whole future is under threat of being abandoned. In the Rocky Mountains, the battle over water in the Colorado river is catching all the headlines with Lake Meade dropping at an alarming rate. Big cities such as Phoenix are clamoring for more water, irrigation well water has now fallen to the 600-foot deep level. California is building tunnels to divert rivers, all this while the source of their summer water is about to disappear. Apparently forgotten in all of this bickering is the fact that the Hoover Dam and Lake Meade were built for a hydroelectric project and had nothing to do with agriculture. The issue is whether that should be changed to a storage system for City use?

–Aquifer status is best judged by how deep wells have to be drilled versus in the past. As an example, in the lettuce growing Valleys of the West (Colorado, Arizona) well depth has dropped from the 2-300 ft. to the 4-600 ft. depth with 800 ft. the bottom. Science meantime describes it as being 20% lower, but the growers know different with going deeper impractical spelling cessation within the next ten years.[12] (These reports are frequently hard to follow due to aquifers being sloped.) Regardless, more water is being pumped out than replaced with even the huge Ogallala reservoir losing water at an unsustainable rate. And don't even ask the farmers about the salt content of the remaining water for it now dictates what few remaining crops can be grown! (Only now am I remembering something that happened many years ago when drilling wells for cooling towers at Medicine Hat, Alberta, Canada. We hit an underground river with water that was cold, almost pure, and had a flow rate greater than that of the nearby South Saskatchewan River. At the time, none of us realized what a rare event this was. What we did discover was that it had the same analysis as the water from the Columbia ice fields and was heading towards the Cyprus Hills, that interestingly had remained an isolated area surrounded by ice fields during the last ice age. Today, I wonder if that could have been the undiscovered source that fed the Ogallala aquifer?)

–Surface Pond and dam storage from excess rainfall periods remains this Continent's best option to balance its water requirements.

–California, with a population of forty million, has already exhausted its aquifers and will need to rely on imported sources in the future, for being in an earthquake zone area, the construction of dams to store water is impractical.

Fast forward to Colorado, Arizona, and California in the USA. Using a combination of older proven irrigation systems plus pumping from underground sources (aquifers), surface or flood irrigation turned what had been scrubland into some of the most productive areas in the world. Vegetables were king with supply to all of North America their market place. It can be that way again if stored rainwater can replace aquifer supply.

South America:

For a Continent blessed by arguably one of the most prolific and biologically diverse ecosystems in the World, what has happened is a tragedy that is still unfolding and whose end has still to be written. Encompassing some of the worst

---

[12] D. Perrone and S. Jasechko 2017 Environ. Res. Lett. 12 104002

examples of human greed, avarice, cruelty and ignorance to be found on our Planet, and despite some half-hearted efforts that are collapsing as their political situation is deteriorating, that situation may be heading towards our World's worst calamity ever, and that is saying something, for WWI and WWII takes some beating.

In the North, Venezuela is plunging into depths it hasn't seen since the days of Simon Bolivar, if even then, which I doubt. What we are glimpsing today is the birth of a ruthless dictatorship, that to achieve its ends, is prepared to see that country destroyed along with any semblance of liberty, never mind democracy, surviving. As for the poor, they are pouring into their Favelas (shanty towns) in their thousands with conditions there bordering on feral and without the slightest vestige or trace of civilization as we know it. Entry into one by an outsider is said to be equivalent to a death penalty.

To the South, Brazil is only playing with children's games compared to Venezuela, but in their own way are even more destructive for they are destroying the very fabric of their land right down to the jungles, wildlife, and ecosystems upon which even their own population had depended. However, in one thing they excel, their Favela are even larger and if possible, reported to be even more brutal and bestial than Venezuela's. In the interim, law and order has broken down in both countries with developers and the powerful behaving and doing as they please without regard to regulations, laws, or even regard to civilized behavior.

The deforestation of the Amazon Basin's tropical forests is currently following the previous pattern of uncontrolled slash and burn behavior that resulted in fragmented deforestation, such that few large stretches of forest will be left within two or three years. As for tropical lateritic soils, who cares, for it has become a sick joke? Ask the people of the Favelas in Brazil what happened! The trees were removed by slash and burn leaving soils with 2% or less organic matter and only fit for grass and the grazing of cattle, and instead of one head per acre, because of its reduce fertility, needs 5 acres per head! The crime, (for that is what it should be called) was how that clearing was done. When felled, the trees should have been burned to produce charcoal, with that being worked back into the soil.

With such poor soil, only soybeans, which supply most of their own nitrogen, can be grown. In fact, when farmers try and grow other crops on that soil, pathetic yields result, for without humus and its carbon element, neither moisture nor nutrients are retained in the earth that support plants. (The torrential tropical rains wash everything out of the soil.) Just when will they learn the mistakes of slash and burn clearing and the way that results in catastrophe for their soils? The real crime is that the people there know all about what charcoal does to the soil for they have pockets of 'terra preta' as they call it, throughout the Amazon (see carbon section for details). Even sadder is the fact nobody seems to care. As for those trying to help the people of the 'Favela,' after 10 million, and 5 million in Sao Paulo alone, they long ago stopped counting and have left. However there still remains a hard core of people touting the so called 'New Farmland' that was supposed to be Brazil's economic salvation; but the reality is that even fewer people than ever can make a living from that compared to what used to be semi-jungle land, with those people now the citizens of the Favela. With no sewage, utilities or infrastructure, these are the new jungles of S. America!

What more can happen to the Amazon you ask? How about building hundreds of dams throughout the area to divert water to towns and cities? Over 100 dams have already been built with another 100 in the planning stage[13]. That land will now receive less water and become a prime candidate for desertification. As for the fauna and flora of that area, how can the world sit by and watch this desecration of nature take place?

—Water seepage from the vast Amazon Basin is the source for their recently discovered aquifer that stretches into Paraguay, Argentina, and Uruguay. One can't help but wonder if it too will in the end be affected by all this deforestation?

—In the Andes, the high people of that area may soon have to give up their home for without water; their terrace irrigation systems will fail. Further down the mountains and in the Amazon Basin, over one hundred dams have been built to store water for their cities and towns with another hundred in the planning stage. One would be tempted to ask what it might do to the aquifers, but given the chaotic state of the Governments of Venezuela, Brazil, and some others, one is entitled to ask why? Regretably, there are large blocks of the World that can be expected to do nothing and should be listed as non-participants.

The other countries of South America are almost like a breath of fresh air. All unfortunately, suffer to varying degrees from the twin problems of rural poverty and lack of infrastructure though all are making some progress. They all show one thing in common; when was the last time you heard that a country in S. America had a positive balance of payments[14] on trade? And now people wonder why they don't have money to spend on capital projects? Though the Caribbean usually hides under everyone's sails, we have one great photograph of humanity's stupidity; it shows what the difference can be between two styles of regimes; Haiti and the Dominican Republic.

---

[13] MONGABAY report.

[14] World Bank 2016 report

Earth, a Planet in Peril

Haiti forests -V- Dominican Republic Forests.

Courtesy emaze.

On the right you can see the forests of Dom. Rep. and on the left, what remains after the Haitian destroyers have left!

If you think the leaders in Brazil are the craziest people around, think again, for Indonesia, with it also using slash and burn to clear its forests to grow palm oil, almost put them to shame, with their soils now rapidly deteriorating. And so, everyone, this is what lies ahead when we talk about clearing forests to make way for billions of more people. And this before we have talked about what we have done to the wildlife and biodiversity of our planet! (I removed that section due to the horrendous nature of the photographs, a sight not even fit for adults.)

The people of the Sao Paulo Favela bid you farewell!

Middle East

And now to the place where farming all started. Damage to our environment is nothing new. Even our earliest settlers were, in their ignorance, ruining what the Bible said was "A land flowing with milk and honey." In 10,000 BC, and emerging from the last ice age, our race was one of being hunters of wild game, naturally growing fruits and nuts, and fish from the sea, rivers, and lakes. Moving northwards, perhaps from scarcity and population growth pressures, and traveling through Egypt, the people reached the Syrian area by 7,000 BC.

Archeologists discovered the first traces of modern farming, wheat seeds[15] for planting and harvesting. By 5000 BC, those seeds had traveled throughout the Mediterranean including all the islands. The reason for that was not because they had commercial seagoing boats, but rather the battles that were occurring between the farmers and the 'hunter gatherers.' With such extreme differences on how to feed oneself, it was inevitable that these two groups would constantly be at war. One can imagine how the farming communities would keep looking for a place that had natural barriers to protect them from those attackers. In desperation, they would have traveled to those islands using river rafts and early boats for that hazardous journey, taking with them their seeds and goats. And now begins our story of damage and destruction. In those days, much of the Middle East was densely forested and the first settlers had to cut down trees in order to farm. However, if large and of cedar wood, there were plenty of buyers who wanted that lumber for building boats and buildings. If not, it was burned. Next came the planting of the first wheat seeds, selected from the natural but largest heads of grain.

The Timeline of Early Wheat Seed for food.

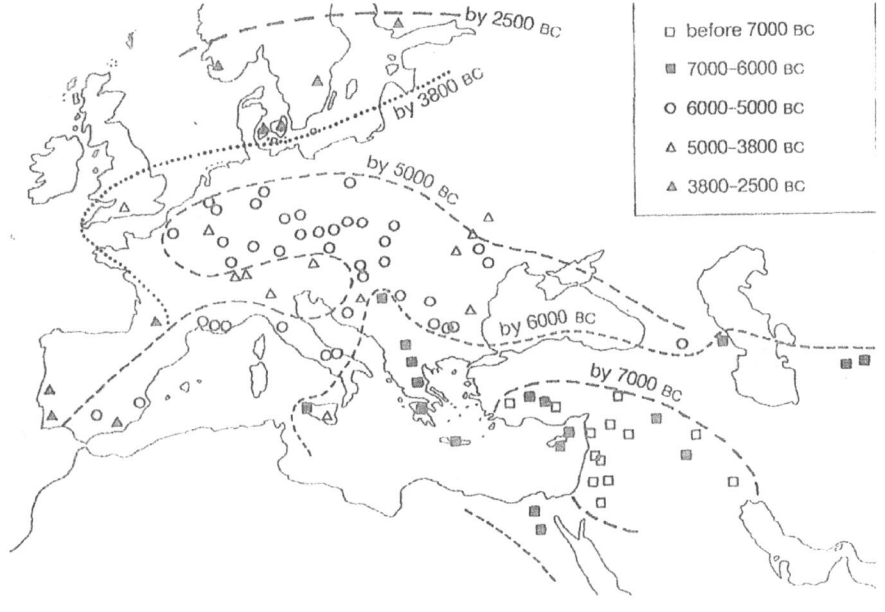

---

[15] Dr. D.C.Johanson, The First Humans, Smithsonian.

Shortly after came the grazing by goats. These are unusual animals in that their teeth are so far forward in their jaws giving them the ability to eat grass down to the bare ground.

Our Goat's Teeth.

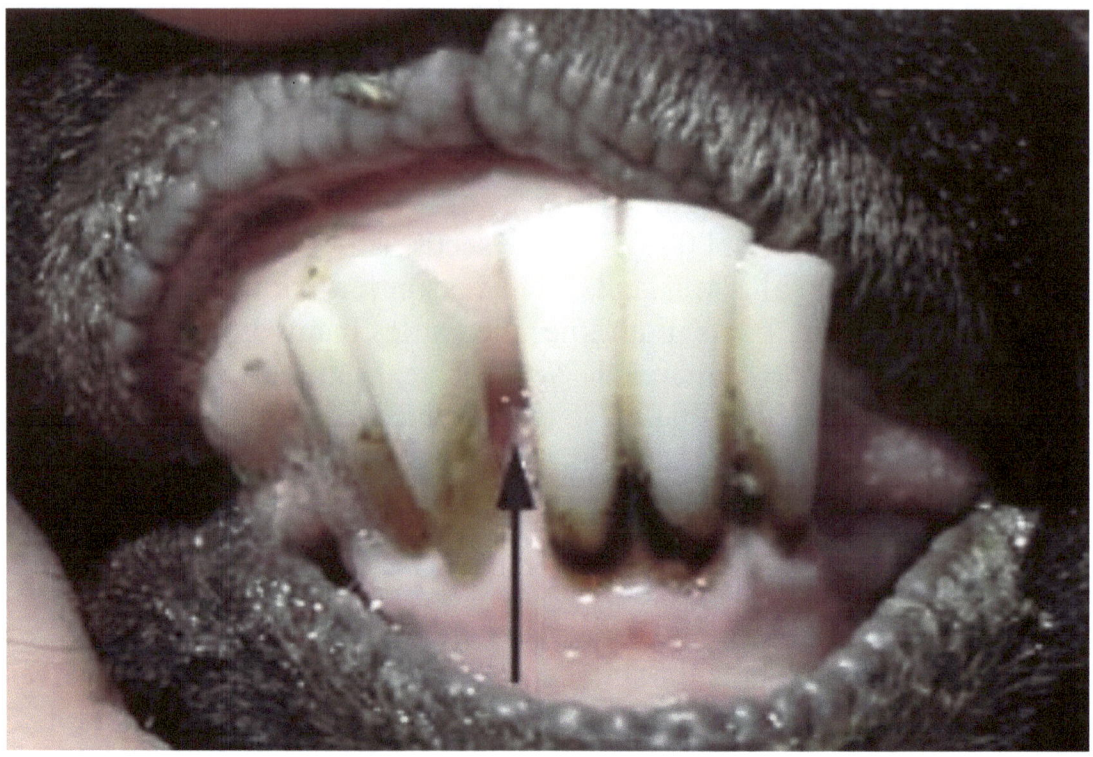

Photo Courtesy of Wikipedia.

Unfortunately, that ability included the grass nodes (crowns) from which the new sprouts grew, and so if overgrazed, would be killed. With a combination of the cutting down of trees, their burning, and with the rudimentary digging of seedbeds for their wheat, the coup-de-gras was the over-grazing and killing of the grass by their goats. With bare ground, and no defense from rain or wind, the land of 'milk and honey' eroded, blowing and washing away it's topsoil into nearby valley's, rivers, lakes, and finally, into the sea. If one can imagine what this did to the tree covered hills and islands of the Mediterranean, whatever soils they had were soon gone. As for the trees, they were all gone with dung now the fuel source. Even for building, stone had now replaced wood as had reeds in shipbuilding. The final curtain for wood came with the bronze age and its demand for charcoal.

The impact on climate and population was negligible for as this land was lost, so too was new land to the north uncovered, for, with climate warming, the glaciers were retreating exposing unspoiled fresh earth or with fresh soil deposited as the ice above melted. Luckily the farmers of those days, from trial and error by the best of them, soon discovered the merits of crop rotation and the benefits of animal manure. Thus, the northern area, and the flatter areas chosen by farmers was largely spared from the depredations of massive soil erosion that had by now permanently devastated the Mediterranean[16]. As for the hunter-gatherers, they were the ones that were fleeing to the north and living in the heavily wooded areas in places like Germany, Poland, Russia, Scandinavia, and Britain.

Before our modern age and our use of fossil fuel, plant and animal life had historically remained in balance, and we might be tempted to think that nothing on the environmental front was happening. Regarding the measurement of CO_2 in the air, that might be true, but don't forget, that was due to the fact the oceans still had lots of capacity to absorb CO_2, and the quantities emitted from wood fires were still relatively small.

Irrigation from lakes and rivers is done by pump to irrigation canals. From there it is again done by pump or tubes to the field. As a consequence, irrigation is restricted to the river's valleys and deltas. Some of these systems are extremely old with the Canal of Joseph in Egypt dating to 2000 BC, the longest and oldest gravity fed system by far, and incredibly still in use today making it the oldest operating system in the World. (And still not recognized by Egypt, or by UNESCO as a 'World Heritage Site'. Adding insult to injury, the younger Qanat system in Iran, which I personally worked on, while also gravity fed, is mostly inoperative due to land salting-up problems and tunnel collapse yet has been awarded that title). Only in the case of high dams have irrigation canals been able to transport water to more distant places, but these are dams built for electricity.

Tradgically, Egypt has come from almost the perfect agricultural system to one of impending doom with nothing yet in sight to mitigate their and/or, their neighbors, bad decision making. Unfortunately, it was not because they changed the Nile River from solely serving agriculture to generating electricity, it was because they did not understand how irrigation must be done in that kind of climate and without their usual annual floods. The problem is salt movement in soils and how that is managed. For non-farmers, salt has to be washed out of the soil into a tile drainage system where it is then drained into evaporation ponds with the salt then later being disposed of safely. In other words, in Egypt, it is still the old flood system with the only difference being the salt build up in the soil ends up in evaporation ponds instead of the Sea, and depending on the crop being grown, may be only done as needed.

At the end of the last 'Ice Age', both the Sahara and Arabian desert had much higher rainfalls than today with major aquifers under the surface. The Sahara's have remained virtually untapped whereas Arabia's has been draied to the point that in 2016 Saudi Arabia was forced to shut down its wheat growing sector (the sixth largest wheat exporter in the World) to preserve what little water was left. As one can imagine, with desert above, there is virtually no chance of that water

---

[16] While farmers are aware of the value of using charcoal as a soil amendment to improve their land, it's use is still only a fraction of what it should be.

reservoir being refilled given today's climate.

While Turkey is relatively well off with water, its land is still very much a casualty of ancient times. Only a program to restore the soil's organic matter and charcoal content can help. Unfortunately that is not the case with Iraq and Iran for, water supply there is in even worse shape than their soils.

Europe

To pretend that more forest clearing could feed a larger population of humans could not be further from the truth. Our ancestors were not stupid. They had already chosen the best land to farm, with the marginal land soon being abandoned and returned to forests. Indeed, the huge part of the current listing of forest land is in the tropics, northern tundra and steppes areas, with virtually everything else being in mountain, swamp or reserved parkland. As for the reforested areas of the world, how many of them are on rich farmland? Not anything of significance with most of those on estates and other private areas. One last thought before leaving this subject. To feed even one year's growth in population would mean that all the world's wine, beer, and alcohol growers would have to convert their land to wheat, food that all people need in the form of bread, pasta, cereal, and biscuits. I left out cake here for it reminds me of the sad story of Marie Antoinette. When she said, "Let them eat cake," she only meant there was plenty of soft wheat to make cake, but no hard wheat to make bread. Talk about a misquote and a lie for political purposes!

By the mid-1600's Europe had lost most of its hardwood forests to support shipbuilding. The killer for them was the 'Iron Age' and the resultant demand for charcoal to smelt that ore. Total deforestation followed with the turning to coal, the only way to keep going. Later in the 1800's, the iron industry had started to move to America, for charcoal was still a hotter fuel than coal, and there were lots of forests there to support that industry. Just to complete the story of deforestation, the softwoods in Northern Europe were also cut down to manufacture preservatives for the wooden ships. The name 'Stockholm Tar' should give us a hint how the whole of Scandinavia and NW Russia became totally deforested and were forced to sell off parts of their countries to make ends meet!

By the late 1800's, the 'Steel Age' and the Bessemer Converter opened the way for coke to become the fuel of choice, with the mass production of steel replacing iron.

The huge change for humanity and its way of life was happening at the same time. First came the steam engine and railroads in 1829 using wood and coal. Next came the discovery of oil and its use to power automobiles and opening up the new fast and easy industries of land, sea, and air travel.

Next, the discovery of electricity and the use of coal to light up and power our world was the game changer. These developments opened the way for humanity to expand such that there appeared to be no limits to how many people could live on Earth. Many things were converging into the present moment which could be called the pinnacle of humanity's climb to civilization, or conversely, the threat of death to most other life on our planet.

73,000 BC. to AD. 1829, our population grew from 300 people to nearly 900 million.

1829 Railroads started.

1859 Oil discovered.

1876 Steel replaced iron.

1879 Electricity and the light bulb invented.

1890-2017 The human population grew from 1 billion to 7.5 billion.

2018-2041 If continued, due to the large child bearing portion of our population, this growth is expected to increase our population to 9 billion in only 24 years, and 10 billion by 2054.

Farming in Europe today is more a case of continuing to optimize per acre yield for there is little new land available with more being lost to urbanization every day. Conversely, the opportunities for Horticulture and Fisheries are seemingly endless.

The Indian Sub-Continent

The Indian Sub-Continent, with just India, Pakistan and Bangladesh, are facing adding 415 million people over the next twenty-five years, never mind struggling to meet today's demands. It is a number that spells a disaster is coming that is almost impossible to imagine, never mind the chaos and lengths to which those people will be forced to go if they want to avoid starvation and death, with war perhaps an attractive alternative. It is hard to ignore the self-congratulations going on among the many pointing out that as education and awareness happens, so too do birth rates decline to the two level, but that is not the problem here, it is the fact there are so many young people of child-bearing age such that even at the two level the population forecasts have been based on that fact.

The conundrum of this continent centers around religion and the status of their holy river; 'The Ganges'. To this day it remains left to its own devices and therein lies the problem. With nearly 800 million people living in it's Valley, and nearly all of their sewage and waste, including their dead, being discarded into that River, the people still bathe, wash and drink that toxic brew. For some, with access to power and pumps, water from the aquifer that lies below has been their savior to this point, but that versus water for crop irrigation and food will be the coming battle.

Earth, a Planet in Peril

The Indian and Himalayan Sub-Continent.

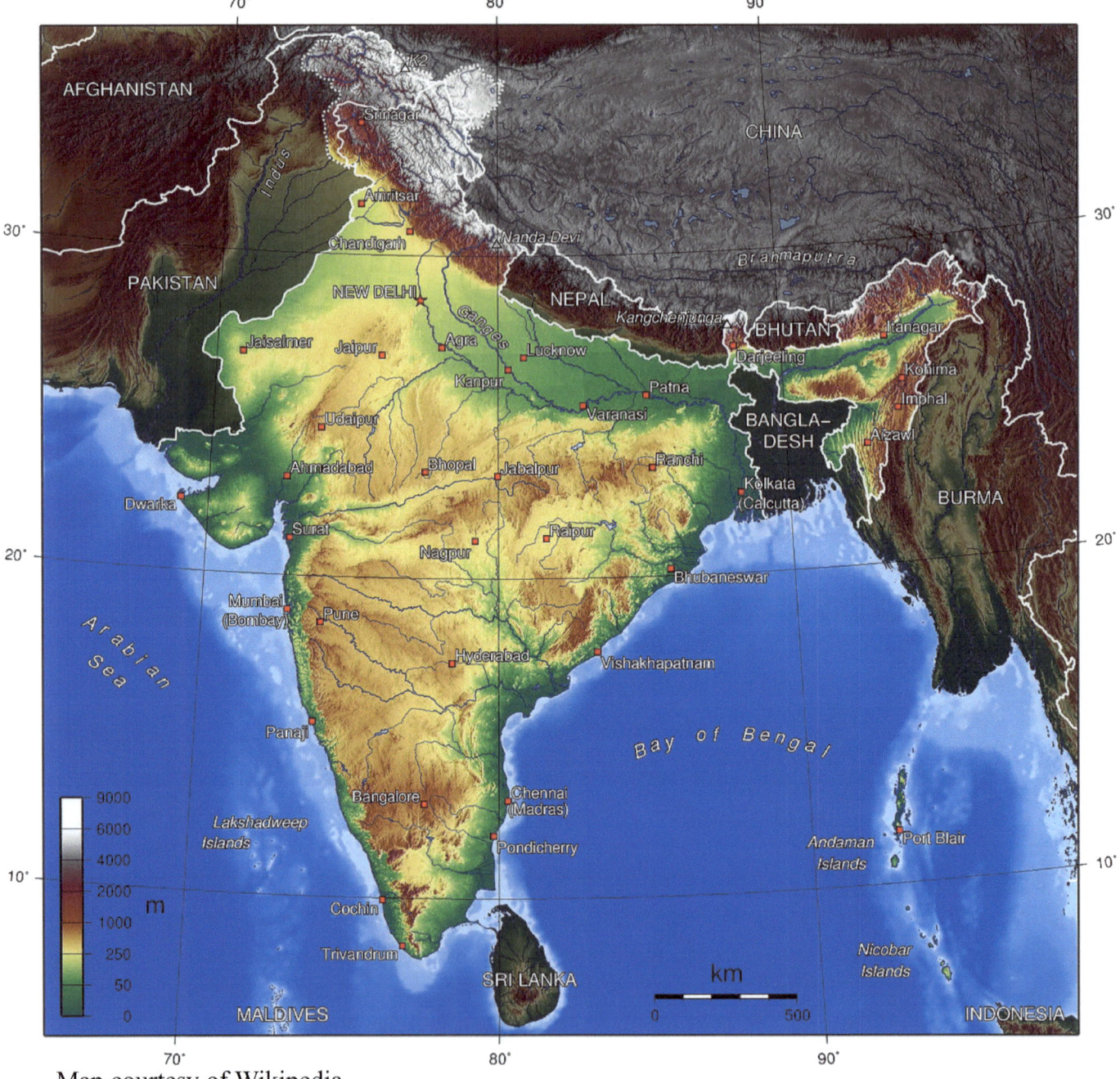

Map courtesy of Wikipedia.

As for the statistic keepers of our World's forest areas, have you ever wondered why they do not keep track of what realistically could be converted to productive farmland? I have no doubt they have, but were afraid to publish that data, for it would once and for all have dispelled any illusions as to what is lying ahead for our world. For example, they show both Bangladesh and Pakistan having nearly 15 million square kilometers each of forest land left. They may have in the mountains, but none is left that could be farmable. As for India, they show over 706 million remaining with all of it in the mountains or the thin soils on top of the 'Deccan Traps'! See that strip of flat land south of the Himalaya Mountains, well, it contains 80 % of the entire population of all three countries! Notice anything else? All the land South of it is mountainous, stony and much of it unsuitable for farming. What else can you see? Why, all of it relies on summer water from the Glaciers in the Himalaya Mountains. But wait, many of them will be gone in the next 5 years with many more gone within 25 years. And we are forecasting an increase in their population???

India and Bangladesh's key Agricultural producing Area

Map courtesy of Wikipedia.

With the Country already bursting at the seams, what a surprise, India's forests are all in the mountains. Indeed, even the World Bank admits that 25% of the World's forest areas fall into that category. As a consequence, everything in India relies on the state of the Ganges River and the fresh water Aquifer that lies beneath it. In combination they feed the water that is key to their irrigation system from which they raise their food. Thus, one can say, as the Ganges goes, so does India and Bangladesh. For that reason, I would like to repeat an earlier section here for it tells us what may happen.

The fresh water source for the Ganges comes from the Gangotri Ice Field Glacier in the Himalaya Mountains.

"There is a huge difference between an 'Ice Field' and a 'Glacier'. The 'Ice Field' is more like a pancake that has been poured over a bumpy surface with the thin parts being on the mountains and the thick parts being in the 'Glacier' valleys. Hence the 'Glacier' probably contains most of the frozen water, but not always and by no means will it melt at the same rate as the 'Ice Field'. Using the analogy of the pancake, the thin parts will melt first and more quickly with the solid ice parts, the 'Glacier', more slowly, for it also may be shaded from the heat of the sun. As an example, in Canada's 'Columbia Ice Field', the area covered has shrunk by one third in 20 years, but that does not mean that the 'Glaciers' have also melted at that rate. In fact, without having access to the ice volume loss data, I suspect it would be 25-50% less dependent on the amount of meltwater in contact with the 'Glacier'. An excellent example of this phenomenon can be found when looking at the 'Ice Field' that supports the 'Gangotri Glacier' in the Himalaya Mountains, and is the prime source of water for the Ganges River of India. There, common sense tells us that the vast 'Ice Field' will melt first, but the main 'Gangotri Glacier' might last twice as long and well into the next Century. The downside of that is the flow rate of ice melt into the River itself would probably, by then, be a quarter or less of present flow rates, if behaving similarly to the one in Canada.'

The Gangotri Glacier, source of the Ganges River of India

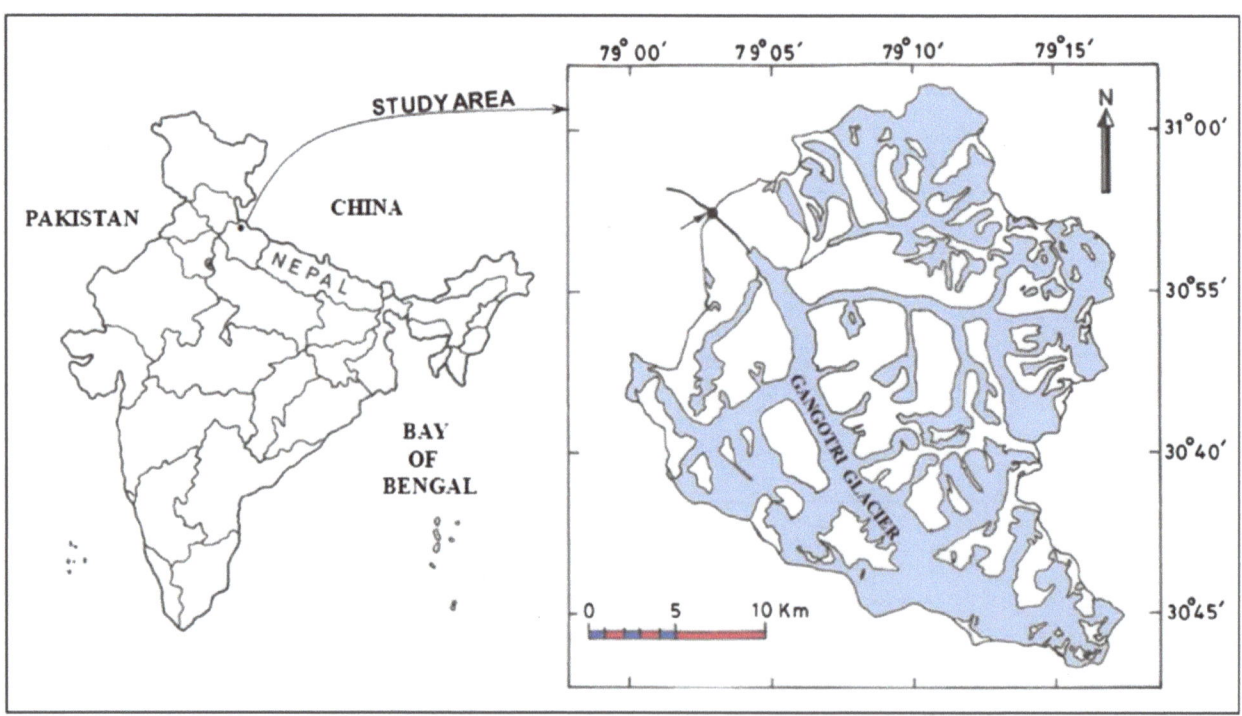

Courtesy of Rakesh Kumar, Naresh Kumar, Jatin Malhotra, and Manohar Arora, Assessment of Sediment in Himalayan glacier., Hydrology Research April, 2014.

In other words, there are two issues about the Ganges that need to be considered; the expected length of life of the Glacier, and the amount of water being fed into it each year or the 'flow rate'. The good news is that this Glacier may last well into the twenty-first Century. However, the bad news is that long before then the flow rate may drop by as much as 50%.

Now to the River's management, or should we say, lack of management. With almost no sewage treatment and the whole River being treated as one huge waste disposal system, what is going to happen if the River's flow rate drops by 50%? The pollution level, just from the arithmetic will double! A prospect that would spell doom for the people living around it not to mention agriculture and the growing of crops for animal and human consumption. And this before we factor in that coming explosion in population. As for what is going to happen in Bangladesh and the Delta area, we can only pray and wait.

By contrast, Pakistan, with the Indus being fed from the World's largest 'Ice Field' (outside of Greenland and Antarctica), is lucky, for the Indus might survive in its present form for the next 100-200 years.

SE Asia

Does everyone get the message? The Khmer, more than any other people in ancient times showed us how to capture and save rainwater to feed and support a civilization such that no other source of water was needed. Unfortunately, its current rulers are dismantling and abandoning that system in their quest to produce electricity for their voters, and so, they for one, are about to relearn the hard way how to manage rainwater, so that year-round, or at least during the growing season, water is available for irrigation. Adding fish was also a valuable byproduct. So, what should we be doing today? Regrettably, in flat lying areas, land has to be given up in order to provide areas for storing water. However, land without water during crop growing periods is not worth much. Off-setting that surrender of land to water storage is the combined factor of dyke building and excavation which can reduce the loss of land for water storage from 10% to 2.5% if the excavation and dyke height is quadrupled. Regardless, the storage of water during monsoon or high rainfall periods is the only practical alternative to the facing of drought conditions. Even in the USA, this type of water storage can save money in many locations with the Houston area only the latest example. (There is also the potential to use this type of water management to store energy.)

Central Asia & Japan

We hope Japan's leaders continue their tradition of protecting their way of life and their land management system, for not only is it one of the most efficient ecological systems in the world, it is also a system that makes it self-sufficient based on rain-water alone.

On the other hand, China is in the process of stepping back from their somewhat similar system with the story soon to unfold, for the changes coming are to say the least dramatic. The fight, if any, will be about water, and we already know how the first vote has been cast; electrical generation from water at the Three Gorges Dam. However, unlike India, China has a firmer control over water pollution such that their two major rivers should remain healthy despite lower flow rates.

Data on China's aquifers, while sparse, indicates that they too may find that source short lived, especially in the NE section of their Country.

The big change in that Country has been from a rural to a City shift in population. It has also entailed a huge shift from an intensive people and livestock driven system of farming to one of modern 'high-yield' crop varieties and mechanization. The 'Achilles Heel' that inevitably follows, of reduced soil organic matter and resultant lower water and nutrient holding capacity, has yet to be felt and, to say the least, will be interesting to watch.

Asia and Glaciers.

Courtesy Meltdown Tibet.co @ Michael Buckley.

This map perhaps demonstrates better than any other how Asia's greatest vulnerability from over-population is to fresh water supply.

Oceania

New Zealand, in terms of looking after it's environment runs a close second to Japan. In many ways it now has one of the best environments in our World. Perhaps the answer lies in the fact that these two Island Nations have had to rely on their own resources and to make sure they are both renewable and self-supporting.

Australia, despite having perhaps the largest aquifer in the World (under the East Central part of their Country), has its problems, for the main crop growing area is in the SE Coastal region. In the meantime, growing conditions are dictated by a sometimes erratic rainfall pattern. With much of the rest of that Continent being desert, rainwater storage appears to be the only solution.

New Zealand's Pride and Joy, the Franz Joseph Glacier.

Wikipedia. One of New Zealand's most spectacular Glaciers

Africa

I left the worst to last! In Africa, with so many small countries, the huge forecast numbers for population growth do not jump out like the big three of the Indian sub-continent. If added up, how many realize the population growth there is going to be nearly 700 million. And that, in a Continent that is already under extreme stress. Realistically, if we add in the Middle East countries, that number rises to 800 million. We are talking about adding 36 million people a year between now and 2050 in an area unable to support what they have today. Spelling out the details of what might happen we leave to others, for this is a book about how to fix and prevent problems. However, one fact should be pointed out to the leaders of Europe; most of these people are from what used to be forested areas and are 'Hunter-Gatherers' rather than from farming communities. They are used to killing and treating all land as being without ownership. Something alien to modern Europeans, but more importantly, these people are used to moving to somewhere else when there is no longer food to be found and will kill to get it! The ultimate tragedy is that many of these people are from sophisticated Tribes rich in heritage but are doomed by the killers among them.

It is tempting to describe next the picture of Central Africa, but it has no Sao Paulo, and without that, the poor are condemned to remain in their villages or to try and escape to distant countries. In fact, last year a friend of mine described a Country he visited as having a fleet of modern ambulances, with only one problem, there were no hospitals or doctors in the entire Country! Almost beyond thinking about was where they would be dropping off their customers, none of which were heard from again.

Where to begin is the challenge. In the East, we find the worst example of what humanity can do to our environment, not just deforestation, but the loss of virtually all wildlife and their soil from erosion. The Island of Madagascar, in what was until a few years ago one of the World's great and unique ecosystems, and home to thousands of now dying species, is in the process of being totally destroyed with its inhabitants sliding into the scars and detritus of their broken lives. On the other side of that Continent is a land where the original 'Hunter-Gatherers' had formed large families or tribes. Finally abandoned and left to their own devices, the final insult was that they were now a group of people aligned with what had been conquered rather than their original family or tribal members.

The Countries

. Further compounding these difficulties, Africa's climate dictated that there would be three wide bands of vegetation which was desert (The Sudan), dense Jungle, and then again desert (The Kalahari). None of which were exactly easy to clear for the more traditional way of raising food…farms.

Earth, a Planet in Peril

The Rivers of Africa

Map courtesy of Lizardpoint.com @ Lizard Point Quizzes.

    Nigeria, although most there don't know it, is already a powerhouse. Potentially, it has an agricultural base second to none in that Continent and with a large river system could easily build a vast water storage system.

The Aquifers of Africa

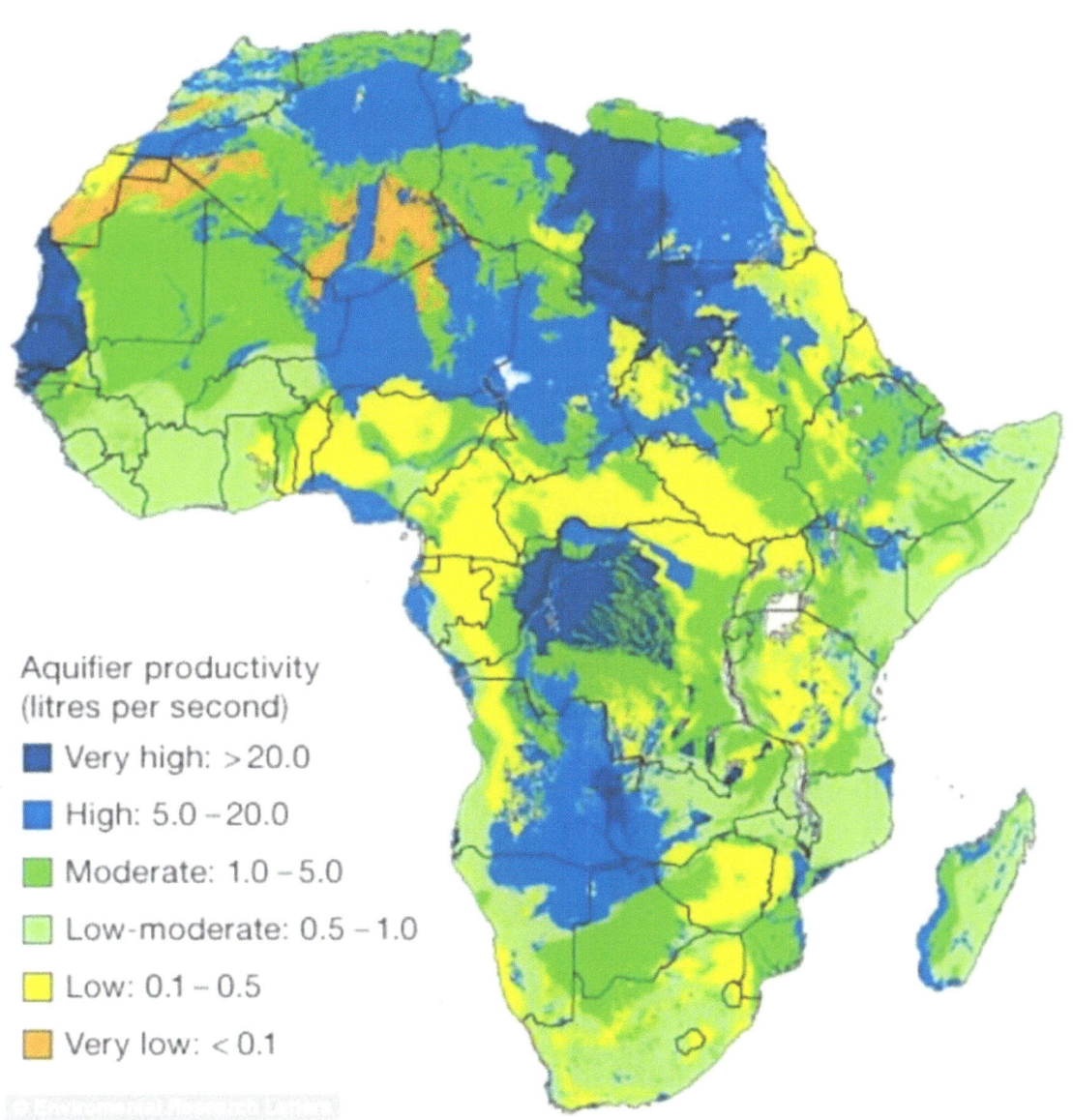

As for the source of capital, if their oil and gas industry revenue was directed to that country, who knows how quickly things could change. It is not by chance that ISIS has chosen that Country as one of its chief destabilization targets.

Zimbabwe and South Africa would make a formidable pairing not to mention the other countries in the Trans-Africa Union. Almost the mirror image of Nigeria, just replace oil and gas with minerals, then you would have two World beaters.

Switching back to NE Africa, just for a moment imagine what might happen if the Arabian Peninsula and the Saharan countries joined that group!

To me, the above would change Africa into a World Leader overnight, not to mention giving it the resources and strength to solve its own problems, for that is what needs to happen.

The Sea

Several things have been happening to our Seas and Lakes that even in the USA we have been unable to cope with or solve. Unfortunately, they are the same three weaknesses that seem to dog our species around the World. Greed, ignorance, and politics.

The story of the Chesapeake Bay is one of which the USA should be ashamed, especially so in that it lies at the steps of Congress and the White House, so to speak, although few go into that water.

When the early settlers dropped their fishing nets into the Bay, only complaints were heard. They were so full of fish, and the nets so heavy, they could not be pulled back into the boats. Cutting their nets and releasing the fish was their only solution.

Hardly was that solved when another complaint was heard. They kept snagging and tearing their nets on reefs made up of millions of dead oyster shells.

Have you ever heard of the clamshell bucket used in excavation? Well, it was named after the device invented to destroy those same oyster reefs. Within fifty years, the Bay was becoming cloudy and fishing such hard work that the fishermen now had to work half days to make a living. Little understood at the time was that without those millions of oysters, no longer would the Chesapeake Bay's water be filtered of its silt and from now on would be cloudier. Also, without those reefs, the shellfish population was heading for eventual extinction, for oysters do not do well on mudflats. Next went up a dam at the entrance to the Susquehanna River. No longer would river fish exit to lay eggs up its channel. Concurrent was the growth of the new Government site of Washington, D.C. on the Potomac River. Oops, to save money, Washington combined the rain and human waste sewers into one system. It worked, save for the fact that when it rained heavily, their system could no longer handle both at once and so everything, including raw sewage, was then released into the Potomac, ending up of course in the Chesapeake Bay. Concurrent with this was the complete deforestation of the area for steel manufacturing and farmland, and later for urban development. Perhaps this last was the final tipping point, but regardless, the combination of all these actions spelled doom for the Bay, becoming an almost lifeless body of water that we find to this day. Mounting clean-up programs, but god forbid, not stopping attempts to catch the last fish in the Bay, sixty years later it is both cloudy and bereft of domestic oysters and any meaningful harvestable fish. That dam is still in place as is the Washington D.C. sewage, or should we say flawed sewage system.

With that as an introduction, things can only get better you say? Think again, for the world has been treating the sea as if it were one giant sewer, by the adage of; 'if you can't see it, then it's okay.' The combination of habitat damage and overfishing has brought the World's fishing industry to the brink of collapse. Again though, there is a hidden killer at work that may doom all Ocean life to extinction. The skeletons of sea life are based on a mixture of bone, cartilage, and calcium. As carbon dioxide is absorbed into sea water, it forms carbonic acid, a mild but never the less acidic compound, which on contact with calcium forms calcium carbonate or limestone, a highly stable product. Without free calcium, much of sea-life struggles to build their shells and protective structures so necessary to their survival[17]. Particularly sensitive are certain forms of plankton and the formation of coral reefs. While studies[i] to date have not yet defined the level of acidification that causes the death of a species, they have observed a steady deterioration at current levels. Of greater concern is the domino effect that this has on the food chain.

*****

---

[17] IISA study; Ligia Azevedo, 2015.

8. THE STORY OF ENERGY

(All data from US Govt. sources.)

The Age of Wood.

For early humans, trees and wood were their friends providing a haven to climb for safety, burning for warmth and cooking, material for building shelter, boats that float, and weapons for hunting. Later, some other uses were discovered. The sap or oil from some trees had special properties providing pleasant aromas, cleansing ability (potash or ash for making soap), tars for waterproofing and protecting ships, and even the left-over ash and charcoal from fires to improve tropical soils. But the greatest discovery of all was to do with how wood burns.

First, there comes the low-temperature stage, or ignition phase, when wood starts to burn. After that, and as temperature rises, the wood burns more fiercely turning into charcoal. At this stage, no carbon dioxide has been released as it is still in the form of charcoal, that black stuff you see on the fire that is starting to glow. Believe it or not, that glow is the indicator that the charcoal is starting to burn, and the temperature is starting to escalate in a major way, coincident with the release of carbon dioxide into the atmosphere. And that too was the secret for the 'Bronze Age' for they used charcoal to get enough heat to smelt that ore into metal. The discovery of the 'Iron Age' was the killer. Discovering that charcoal would also melt those ores into iron and steel, what followed was an environmental disaster of epic proportions[18]. If one can believe it, the whole of western Europe was deforested with the clue being that even today, it is a rare event if a tree older than three hundred and fifty years is discovered. The story of the Armada and the other immense fleets of Europe that took the last of the oak trees to build them only compounded what was about to happen. However, the fleet's ravenous appetite for wood does not stop there. Ever heard of Stockholm Tar? It was made from softwood and was used to waterproof and protect ships from wood boring creatures. With Sweden and Russia, the main sources of supply, they too soon faced complete deforestation, with Sweden selling and buying parts of their Country to raise money as the fortunes of those industries rose and fell. And that, ladies and gentlemen, was how America's iron and steel industries got their start, for there were vast forests here that were about to be cut down to make charcoal. At that moment in history, with Europe deforested, one could say humanity had reached the point where the world could no longer support their demands. In other words, when there are virtually no trees left, the World is in crisis. In some ways, America was Europe's savior, with South America, Africa, and Asia still untapped, but shortly also facing the Age of European Colonialism and deforestation.

---

[18]. The people of those times were so ashamed of what they had done that they wanted to forget it, such that even today that story is hard to find.

During these early days, in a few places such as near Jerusalem, open wells of seeping petroleum tars were known and greatly valued for their many properties. Interestingly, the Egyptian Pharaohs owned that property with Cleopatra and King Herod both asking Mark Antony to cede it to them (Herod was the winner, but not until Octavian defeated Cleopatra). As for products such as coal, written records show that surface deposits were known, but with plentiful wood supplies were little used.

It was copper that doomed the Middle East into becoming the first deforested area on our Planet. Already under pressure from the demands for wood from treeless countries such as Egypt, copper, and then copper and tin (bronze), were the killers.

The 1800's can claim to be the great age of invention for modern humanity with the list of discoveries and inventions seemingly endless with the discovery in 1862 of oil in Pennsylvania the icing on that cake.

The Age of Fossil Fuels.

Geologically, the source of nearly all petroleum products can be traced back to the Carboniferous era. And here we find oil, natural gas, shale, and coal. This was a time when the World's atmosphere was warmer and had a much higher level of carbon dioxide than today. With plants and algae growing rampant under those conditions, and with massive amounts of organic material deposited in layers over millions of years, this material was converted into the products we find today. To put this into perspective, what we were getting ready to do was to mine products that had been laid down over a period of seventy million years and some three hundred million years in the past, and suddenly, transporting them into our modern age. To complete this tale, we are then threatening to use it all up in less than one thousand years! Little did we realize at the time how that was freeing up humanity to embark on one of the biggest population expansions in Earth's history, nor the fact that we were going to be burning all that stuff and essentially moving carbon dioxide back from the carboniferous age into ours!

The second observation is that these fossil fuel products fall into two types of business. The first is 'The Moving Around Business' and is the exclusive provenance of oil. The second is 'The Static or Stationary Business', which is supplied by fixed energy sources whose cost makes them 'non-transportable' and include coal, natural gas, shale, nuclear, hydroelectric, and renewables. Only ethanol from corn, and at that time, a few experimental electrical cars were able to cross that gap. That divide explains why oil is so profitable, while coal and the other non-transportable products are much lower in BTU value, and with few alternative markets, cheaper.

Earth, a Planet in Peril

With the World now alerted to the dangers of carbon dioxide build-up in our atmosphere, the finger was pointed at the burning of fossil fuels as being the main culprit with the USA responsible for half of the World's emissions of that product.

Not many are aware that Carbon comes in several forms (isotopes). When exposed to the sun, carbon converts to what they call C14. When buried in the earth and out of sunlight for over 60,000 years, it reverts to its original form of C12. With most hydrocarbons being laid down in the Carboniferous period some 300 million years ago and being buried since that time, it is composed of C12. Conversely, with agriculture and human activity responsible for the rest and using surface hydrocarbons, that product is made-up of C14. Thus, by measuring the ratio of the two types in our atmosphere, we can determine how much carbon dioxide has come from the burning of oil, coal, shale, and natural gas versus all other sources such as agriculture and animal life. That calculation gets a little tricky[19], for C14 decays over time to C12 and that adjustment has to be made. However, and despite that complication, the difference in the two sources allows us to make some fairly accurate measurements on both what has been the source of past CO_2 emissions and what would be the impact of reducing it. Luckily that work has already been done by the EIA among others with fossil fuel CO_2 the source currently estimated to be accounting for approximately 80% of the annual build up, with agriculture, human activity, and deforestation causing most of the balance. In 2015, the EPA estimated the fossil fuel split by industry to be; 29% for electrical generation, 27% for transportation, and 21% for industrial uses. A bit of history here might help explain what happened next.

In less than twenty years, the USA went from being the industrial powerhouse that won World War Two into being a rust belt with shuttered plants and empty buildings at every turn. The USA was also moving from being the Country that discovered and supplied this miracle of abundant and cheap energy to the World, into being an importer. This, despite being a Country with huge reserves of the several kinds used today. How was this possible?

It is all about what we have done to ourselves, with few seeming to know or care about the reasons, but that fact is threatening to bring down the house of cards. The illusion was that we were the wealthiest country in the World while the fact was we were hemorrhaging to death. Why? With plants and industries gone, we had no choice left but to turn to imports, racking up nothing but Trade deficits over the years, that deficit finally has reached the insane number of twenty trillion dollars[20] and counting, with our nation facing bankruptcy if it continues.

Remember the worry we had about pollution in general? Well, in our wisdom Congress set up the EPA (Environmental Protection Agency), which is a study in

---

[19] PNAS impact of fossil fuel emissions. Heather D Graven.

[20] Bumbling, Fumbling, Stumbling in the Dark, by Fred Graham-Yooll.

itself of how good intentions can be destroyed by bad laws. Our Law Makers, realizing they were about to set up an Agency that was going to pass unpopular regulations, and in particular against some of their biggest campaign contributors, decided to pass a highly unusual Law. They delegated to the EPA the power to set both regulations and the power to enforce them. All without having to return to Congress for that approval. The analogy to Pontius Pilot washing his hands of the death sentence of Jesus Christ comes to mind! With little regard for either consequences or coordination with others, the almost evangelical enthusiasm of these officials was frightening. Setting ever stiffer regulations for some of our most basic industries to reach or else shut down, and with no steps to help industry finance the enormous cost of new mills and plants, it was not long before those industries began to look for alternative supply sources, and worse yet, to simply close and move their operations to other countries.

Almost obscene was what happened next. The World Bank with USA loans and support helped finance countries like S. Korea, who were more than willing to set up steel plants providing ownership remained in their hands. It was not long before China, with self-financing, followed suit. The saddest part now followed. Many of those new overseas plants were worse polluters than those in the USA. Steel plants started shutting down throughout the USA, such that today, all that remains are recycling and specialty units. And now comes the absolute imbecility of what happens when basic industries leave, and something to which no attention had been paid. Who were the big users of our steel? Shipbuilding! Automobiles! Pipelines! Rail and rail equipment! Construction! Within only a few years, S. Korea, China, and Europe became the centers for those industries with ours closing or moving overseas in droves. Even the supporting industries were joining the exodus. Worst of all, no one was brought to task for this mess. It was not just about the Steel Industry; it was about all those thousands of other businesses that depended on them with that domino effect continuing to this day. To relieve the doom and gloom of this tale, let us leave it on the hilarious note of what happened in 2015. Never seeming to give up on passing new regulations, for that was their 'Mantra,' the EPA turned their attention to cleaning up pollution from ships. Easy, they thought, just pass a rule they have to switch from Bunker Fuel, containing up to 3.5% sulfur, to Fuel Oil at 0.1% sulfur. The fact that this applied to all US ships anywhere in the world did not seem to bother them in the slightest, whereas it only applied to foreign owned ships in American controlled waters (for them, easy to comply with, for ships have many fuel tanks and dedicating only one to fuel oil was only a minor problem). The fact that the cost of 0.1% fuel oil was double that of Bunker Fuel was of no concern to the EPA. Normally, such a move would have been catastrophic for such a large US industry, so why was there no public outcry? Our Ocean shipping industry was no more! Of the few left, most were back-up vessels needed to move troops and supplies for the military and a handful for shipments to Hawaii, etc. Where had our 1,000-strong fleet of ships gone? They had moved overseas and were now operating from Foreign Flag Countries. As an aside, the United Nations have also got into the act via their IMO organization. They too have recently mandated that all ships have to comply with these same EPA regulations by 2,020. However, such an uproar has occurred over that cost and there being not enough refining capacity to supply that product in time, the IMO is already talking about delaying that deadline until 2025 or later. However, as this does not apply to non-member United Nation Countries, we wonder where all these ships will finally move? Who guessed correctly? To the non-UN member Foreign Flag countries! Has anyone noticed the outreach in this story? The EPA is now passing regulations that apply to USA companies operating outside the jurisdiction of the USA! Any comments? I remember the day the USA decided to change the way oil prices were set. Up until then, the oil Industry had been the one that set prices depending on market conditions. In times of shortages, prices shot up and when in surplus, down. The recognition was always there that it took many years to find and produce oil and so the lag time effect was usually in the ten-year range.

The new price setting change was mandated to be set by the Commodity Market, with day traders to set prices as market conditions dictated. To a trader, being long and short on something as huge as oil was like a 'gift from the gods' and an opportunity to make a fortune. To a trader, being long or short is all that matters with the time it takes to bring on new production of no concern. Indeed, the longest horizon by even those prognosticators rarely went forward more than a handful of years. Guess what happened next? If there was more production capacity than needed, the price dropped below the cost of bringing new capacity on stream. Producers facing those economics shelved plans for exploration until the markets changed. Overseas, the big producing countries were outraged over what was happening and decided to set up their own pricing system; A Cartel! Thus, OPEC was born, and we all know what that did with the USA stopping exploration and becoming a major importer. Not until the great financial collapse of 2008 did the price of oil escape the Cartel's clutches, with the costs during that time to the USA consumer and country (from escalating trade deficits), running into the trillions of dollars range.

Classification of Energy by End Use

Transportation Use	Static or Fixed Use
Oil	Coal & Natural Gas
Tar Sands	Uranium
Shale Oil	Hydroelectric
Ethanol from Corn	Wood
	Solar, Wind, Geothermal

The first thing one notices is that the Oil boys leave the Static markets alone, for with coal by far the cheapest source, it keeps all the other boys at a low price. Interestingly, on the coal side, in a free market burning value is what all the others have to come back to if they are to grow. The second thing that jumps out is that virtually none of the boys in the 'Static Group' can cross over to the 'Transportation Group'. Only in recent years has a crack appeared in the form of hybrid or electrically driven cars, but without new battery technology, nothing as yet for them to worry about.

Classification of Energy by Supply, Cost, Pollutants, and Practicality.

Product	Reserve Size	Cost	Environment	Practicality
Oil/tar/shale	30 Year Max*	Moderate	CO_2 Crisis	Cut Back
Ethanol	Needed for food	Higher	CO_2 Crisis	Cut Back
Coal	100 years	Low	CO_2 Crisis	Must Cut Back
Natural gas	30 Years Max.	Low	CO_2 Crisis	Last to Cut back
Uranium	100 Years	High	Best, but Risk	Water supply
Hydro	Glacier Melt	Low	River/Fish death	Limits reached?
Wood	Deforestation	Low	CO_2 Crisis	Second Worst
Solar	Endless	High	The Best	Sun/Batteries
Wind	Endless	High	Bird Damage	Wind/Batteries
Geothermal	Endless	High	The Best	Most reliable

* Both Shale and Tar Sand oil reserves could last 100 years or longer, and depending on regulations, even oil may be available. However, that assumes there is no better use for the natural gas up there than to use it for heating those tar sands essentially converting natural gas into heavy oil.

Any person in the energy business would now be asking themselves the key question of what products should I be concentrating? For myself, at the time, I found it a strange question for it fell into what I called the 'Gee Whiz' category, for it really meant you were not sure what you were doing in the first place. The consequence of trying to answer that question was to find yourself in a 'Blowing with the Wind' business, where; depending on what you discovered and where, you could find yourself blowing back and forward in different products, different end businesses, and horrors of all horrors, different skill requirements. The answer for many companies was to become a conglomerate and to cover all bases for one didn't know what the future held. So, let's take a look at what was in front of them.

Earth, a Planet in Peril

The cost of exploration and conversion to usable products and to where they are needed are huge as a percentage of their value. The key test for all sources is to check the economics against that set by oil, for that is the price determinant.

Now guess which market is the higher cost? The transportation sectors.

With a plethora of taxes and other regulations distorting the economics, we concentrated instead on the decisions that humanity will have to make.

-The size and availability of the world's reserves.

-The cost of production.

-The environmental consequences.

-The replacement options.

-The practicality of the plan.

Now before we begin, let's go back to our list and reorganize it to match the above questions to the energy sources.

The Energy Sources

In assessing the production reliability of these different sources, only two are in the erratic supply category, Solar and Wind. Thus, either those two have to invest in overcoming those deficiencies by investing in storage systems, or else the rest of the industry has to work around those problems by becoming more flexible. The problem in a nutshell is the time it takes to start up and shut down the burning furnaces that are used to run the generating facilities with most taking several hours or days to do so with flexibility the issue.

Given the above requirements the efficiency ranking would be:

Ranking Supply Reliability and CO_2 Emission Level

PRODUCT	TRANSPORTATION	STATIC USE
Geothermal	# 1 Tie with batteries	# 1 Tie With batteries
Solar	# 1 Tie with batteries	# 1 Tie With batteries
Wind	# 1 Tie with batteries	# 1 Tie With batteries
Uranium	# 4 Limited Water	# 4 Limited Water
Natural gas	# 5 30-year limit	# 5 Short term only
Oil/tar/shale	# 6 30-year limit	# 6 Impractical
Ethanol	# 7 5-year limit	# 6 Tie
Wood	# 8 Impractical	# 6 Tie
Hydro	# 8 Tie	# 6 Tie

| Wood | # 8   Tie | # 6   Tie |
| Coal | # 8   Tie | # 11 Highest pollution |

There are some quandaries here that we have to get rid of first. No matter what happens, certain types of hydrocarbons will be needed for the foreseeable future to operate ships, planes, equipment with high energy demands, and for chemical, farming, medicinal, and other uses. This infers that at some stage oil, natural gas, and even coal, will be allocated to these uses. Assuming this is done, we are now able to concentrate on the two main uses of energy, transportation and the generation of electricity.

If we are to be successful in reducing carbon dioxide emissions to the levels required, we are talking about replacing 80% of all fossil fuel use in America by renewable energy sources. This means converting a large proportion of electrical generation and transportation to renewables, for not all uses can be converted given today's technology. However, with these two uses so dissimilar, we have chosen to handle them separately. But first to renewables.

Renewable Energy Sources

The cost of replacing hydrocarbon sources with wind, solar, and the necessary storage capability remains formidable as of 2017. Nuclear will be lucky to remain flat, whereas hydroelectric is forecast to decline as water shortages continue in the West coupled with the continuing removal of dams for ecological reasons. As for ethanol, that source is no better than natural gas with the later more economical and forecast to remain flat. The bottom line is therefore how much do we want to encourage renewables, and to pay for that program by offsetting those costs by applying penalties to the hydrocarbon sector? Most of these questions have been answered for us by the EIA and others[21]. When considering that step, we have to know two things; what are the best and worst sources of CO_2 pollution, and second, what are the best and lowest new sources of CO_2 pollution. Normally I would take the EIA data unchanged as that group is hard to beat for their excellence and expertise. However, let's look at the next table for it is essential we factor in the true comparisons. To do that we have to add the raw material, transportation and conversion emissions to make these products. Likewise, with renewables we have to add the emissions from their manufacture, and the batteries to increase their percent utilization such that the reliability factor matches that of coal and natural gas units.

---

[21] Global Biochem Cycles 10. Andres, Maryland, Fung, Mathews, 1996.

USA CO$_2$ Emissions (EAI 2/2/2016 Data).

ELECTRIC	CO$_2$/Million Btu	Adjusted
Coal to Electric	210	260
Natural Gas to Electric	117	135
Nuclear to Electric		50
Hydro to Electric		30
Solar to Electric		3
Above +Batteries		6
Wind to Electric		2
Above + Batteries		5
TRANSPORTATION		
Ground Fuels	157-161	187-191
Aircraft Fuels	152-156	182-186

The effect of the above adjustments is to emphasize the importance of replacing fossil fuel use with renewable fuel

The next table, on the capital costs of commercially sized units, we have looked at three cases of taxes and incentives that might be used to encourage investment in the best choices of renewables to replace fossil fuels. Again, our Federal Agency numbers were used;

USA Fossil Fuel Tax Disincentive Program to be Applied to Renewable Energy Investment.

Electric	Cost *($/kW)	Incentives	Net Cost ($/kW)
Solar to Electric	2,534		2,534
Above + Batteries	5347-1,600		3,747
Wind to Electric	1877		1,877
Above + Batteries	4411-1321		3,090
INCENTIVES (1)	@$1/gallon	@$2/gallon	@$3/gallon
Revenue/year	$185 million	$370 million	$555 million
Private Investment (2)	$555 million	$1,110 million	$1,665 million
Years to Replace (3)	52	24	16

* EAI data. Cost less 30% utilization credit.

(1) 2016 EIA consumption of gasoline and diesel and a special tax on fossil fuel use to raise money for investment in renewable energy. If tax also applied to electricity, above numbers would be halved, i.e. $0.50. $1.00, and $1.50.

(2) Assumes 100% private investment with 25% tax credit or fast write-off on both Federal and State taxes. In the event State does not follow Federal allowances assume Federal payments to States would be cut back elsewhere to balance.

(3) At 90 % as heavy vehicle use still reliant on diesel.

In 2016, the USA renewable energy business spent $44 billion on new investments and received $11 billion in tax credit deductions. However, unless renewed, those credits will be phased out by 2,019.

Another huge issue for this industry is the unreliability of supply as wind varies, and for solar, as night approaches. For example, in 2015 wind electrical capacity was 6.7% but produced only 4.7% of production for a 30% shortfall. Likewise, solar electrical capacity was 2% but produced only 0.9% of production for a 55% shortfall, which after factoring in a night of nine hours, almost closes that gap. What most of us fail to appreciate is how our electrical grid works. When we turn on a light switch, we expect lights or appliances to turn on instantly, and in fact they do. This is only made possible by having the more stable electrical generating capacity running on a stand-by basis to meet that demand. Conversely if the source of supply is unreliable, so too do reliable sources have to increase their production to compensate for that deficiency. (Spreading the cost of running that stand by capacity among all producers.) The unintended consequence of that policy is to remove the incentive for unreliable sources to install electrical storage capacity to compensate, for it has been covered by our State electrical commissions making each producer share in those cost penalties. Regrettably, the worst polluter, coal fired units, also takes the longest to start up and shut down due to furnace heating time needs such that, in some cases, causes the benefits of renewables being completely offset by units having to be on stand-by.

Another way to look at this dilemma is to compare the cost of storing electricity in huge batteries to that of having stand by electrical capacity to cope with surges in demand and the unreliability of renewables. That too has been calculated, and if the Eos battery cost numbers turn out to be factual, the cost of huge batteries may be as much as 50% cheaper than stand by capacity. Something that would be a 'game changer'. Next, we need to look at the cost of electrical energy versus gasoline and diesel.

When evaluating the benefits and cost of solar and wind, that cost has been compared to the cost of electricity. If compared to the cost of gasoline or diesel in hybrid and electrical vehicles, that payoff is 33% better! In addition, does anyone doubt that we should make that transition if it can be economically justified, for if so, we would solve 'Global Warming' overnight. Yes, we are crazy, for there are other benefit to all this. The increase in jobs from domestically produced renewable plant and equipment. The increased taxes from added GNP. And last, but not least, the reduction in petroleum imports as domestic renewable capacity increases. We will come to that later, for it should be what tips us all into the 'let's do it now' category! Here is a simple table to show what happens.

Incentive Program Versus Trade Balance of Payments.

Tax Rate/Gal	@$1	@$2	@$3
Revenue/$billions	185	370	555
Net Private capital	555	1110	1665
Years to replace	52	24	16
Gov. Job Taxes*	$0.2 trillion	$0.4 trillion	$0.6 trillion
Gov. GNP Taxes*	$3.5 trillion	$7.0 trillion	$10.5 trillion
Trade Balance*	$0.9 trillion	$1.8 trillion	$2.7 trillion

*Cumulative impact of reducing petroleum imports as domestic renewable capacity increases, and in the case of Government taxes, the amount collected over 16 years.

Line 1 shows three cases of how much money would be raised by taxing gasoline and diesel at these levels.

Line 2 shows the money which would be available to give to the builders of renewable energy facilities at 25% of total cost.

Line 3 shows the balance at 75% to be raised from the private sector.

Line 4 shows the number of years it would take to replace 90% of fossil fuel plants. (Assumes heavy and emergency equipment would still run on diesel.)

Line 5 shows the cumulative payroll taxes the Government would receive.

Line 6 shows the cumulative taxes the Government would receive from GNP growth.

Line 7 (not shown) is what would happen if a one or two-year depreciation rate was given to investors providing they passed on 50% of those savings to their customers, for without having to purchase fossil fuels, the incremental costs of the renewable energy industry would be near zero.

WOULD ANYONE LIKE TO GUESS WHY THE CHINESE ARE INVESTING IN RENEWABLES AT TWICE OUR RATE? AS FOR THE OPPORTUNITY TO SET UP A SELF-FINANCED FUND TO ALLOW

BORROWING TO BUILD THESE PLANTS NOW RATHER THAN LATER, ONE WOULD HAVE TO BE CRAZY NOT TO DO IT. SORRY, I FORGOT WE ARE!

P.S. We forgot to talk about global warming! But wait, by eliminating the use of fossil fuels, that only leaves population growth and deforestation to solve, and that my friends, is up to all of us to decide.

The Electrical Industry

Electrical energy travels down our power lines and into our homes at just a fraction lower than the speed of light. Thus, when we turn on that switch, if we want that light or appliance to work, somewhere an electrical generator must be running. Edison and Westinghouse both struggled over this conundrum with the ultimate decision made that if they wanted to convert potential customers from gas, whale oil, and kerosene to electricity, they would have to accept that waste. For who would accept a day or two's delay while those coal fired boilers heated up enough to drive the turbines that generated that electricity. That decision meant that rather than operating when needed, these generating facilities would operate twenty-four hours per day to supply light to customers that only needed it for twelve hours in offices and workrooms, then in the evening and night, perhaps for six hours. The next idea was one of security. Replacing street gaslights with electric ones over larger areas helped to balance demand. Over the years, campaigns began to use this idle night-time capacity by offering electricity for sale during these 'off peak' periods at deep discounts from normal rates. The sight of office lights on all night became common as did neon advertising. While not exact, these additions had the effect of requiring some new additions to generating capacity. Seemingly out of nowhere came the age of appliances: radios, gramophones, TVs, water and home heaters, washers and dryers, dishwashers, stovetops, refrigerators, computers, with the dragon of all needs, air conditioning. Now there were 'peak needs' on top of 'peak needs' that neither Edison nor Westinghouse could have imagined, but Tesla solved.

In parallel with the above, the generating side of the electrical industry had evolved from small local generating facilities to huge nearby facilities running on coal on a twenty-four hours basis. Smaller facilities were next added that could be started up more quickly to meet peak periods of demand. With start-up speed all important, these sources of fuel ranged from natural gas (ranging from six hours to twelve for start-up) to jet engines running on kerosene (almost instant). Meanwhile, in an attempt to further balance needs, long distance transmission lines were built linking the regional USA grids together with Canada as a key partner, for they, in turn, had large amounts of hydroelectric power available for sale.

Next on the scene came nuclear-powered generators. Taking several days for start-up and shut-down to occur, this was wasteful, and so they became base-load electrical providers.

We now come to the puzzle that renewable wind sources (intermittent supply), solar (day only and that variable), and geothermal steady), posed for this industry. For this, we have to understand how the individual State electrical agencies were managing the industry.

Their challenge was to ensure that individual base-load producers were not subjected to the up and down needs of consumers, for with two or more days to start-up and shut-down their plants, they would be subjected to greater wear-and-tear and the economic penalties of using fuel to no purpose. Evolved was the idea of power leveling and the sharing of the cost penalties of having to operate for short periods to cover 'peak loads.' Taking several years, and still not fully completed, was the solving of this conundrum of being able to accept these new sources of supply, for it was a new industry whose cost of production was 100 % capital and operating cost, but with zero fuel costs. Fantastic, but who could move over to make room for this fluctuating supply, be it pollution free? Thus, we still see the sight of windmills idle on windy days for if the energy is unneeded at that moment who should be made to shut down and yet not be able to instantly supply energy when needed at the flick of a switch. As for the solar industry, if that trickle of supply becomes a flood, who will be made to suffer? On the demand side, we need to answer the question of whether or not the electrical customer can be asked to reduce some of those 'peak-load' needs, while on the supply side, can those variable new sources be asked to level those out by the use of electrical storage batteries? The ultimate question is of course whether production cost or clean air should be the priority.

Common sense would dictate that while there might be an inconvenience in banning the use of various appliances during 'peak-load' periods, such uses as water-heaters, washers, and dryers could be moved from daytime to nighttime use without hardship. This practice has already been adopted in some areas with some customers and has already demonstrated that benefit, so if adopted nationally, would go a long way to solving the 'peak-load' problem.

Likewise, on the supply side, the electrical industry is already moving towards the construction of utility sized electrical storage systems. These are huge systems[22] with a cost structure at least 50% below that of smaller units, ($150 versus $300 per kWh). In other words, our electrical industry is already starting to adapt to gain the needed flexibility to absorb renewable energy systems and at a cost 50% lower than any household can match.

Another different system is also being tried; water storage. (Pumping water to a higher elevation to run generators at a later date, but not on instant stand by.) Anyone ever heard of the Mississippi Lake Michigan project? This again is to reduce the need for a 'balancing mechanism.'

---

[22] EOS grid battery system.

Considering that this industry and the various States have already encouraged these changes and advances, and not only done that but have ensured that the electrical industry has, with Government incentives, self-financed these efforts, one might wonder if anything more than congratulations needs to be said? The answer is yes, but not from the direction we might expect.

EIA USA Electrical Supply Load Variations.

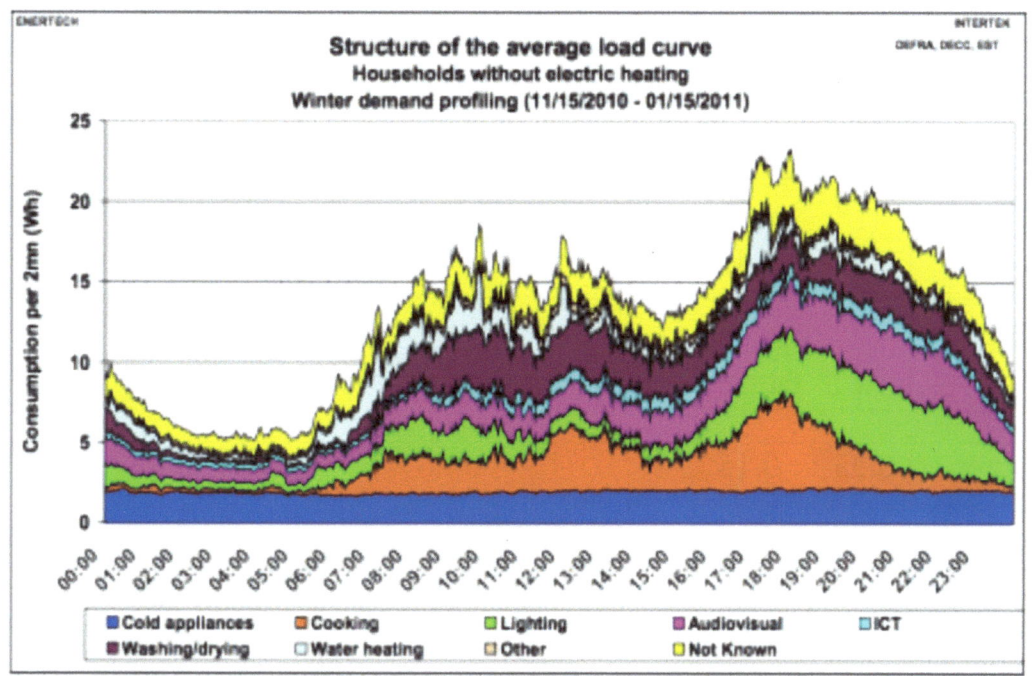

Courtesy of EIA.

Coming back to the issue of 'peak loading,' few appreciate the challenge of supplying that need economically. If one takes a typical city or electrical district, one can expect that 50% of demand will be constant with the other 50% varying with the time of day. There is, however, another dilemma. What happens when vacation, holiday dates, or hot summers come demanding electricity for cooling? This is where larger interconnected districts can help level that demand by 'averaging,' but does not solve the need for substantial stand by capacity. I would submit that until there is extra renewable capacity, the use of natural gas stand by capacity (the most efficient and cleanest fossil fuel source) is the answer.

The Change Ahead

Between the Department of Energy (DOE), the U.S.A. Energy Information Administration (EIA), the U.S.A. Environmental Protection Agency (EPA), and the Lawrence Livermore National Laboratory (LLNL), these organizations prepare annual reports on the USA's energy production, consumption, operating performance, with the latter publishing a flow chart of energy use by source and end use in the USA. It represents a complete breakdown of energy losses throughout the system, thus a magnificent place from which to start. We offer only one caution if one chooses to look at these reports first hand, they are published in two different ways; 'Lower Heating Value' (LHV), and 'Higher Heating Value' (HHV), roughly a 6.5% differential.

Estimated Energy Consumption and Losses in 2016: 97.3 Quads.

(All Data from Lawrence Livermore National Laboratory and EIA)

INPUTS	Gross BTU	Electrical Production	Home Use	Commercial Use	Industrial Use	Transport Use
Solar	0.59	0.34				
Nuclear	8.42	8.42				
Hydro	2.48	2.47			0.01	
Wind	2.11	2.11				
Geothermal	0.23	0.16	0.04	0.03		
Nat. Gas	28.5	10.37	4.54	3.24	9.08	0.795
Coal	14.2	13.0			1.23	
Biomass	4.75	0.51	0.15	0.39	2.28	1.43
Oil	35.9	0.24	0.88	1.02	8.12	25.7
Total	97.3	77.62	1.37	3.857	21.442	27.092

(Numbers may not add due to rounding)

The first measures the BTU content without the loss of heat in moisture or unrecovered steam, whereas the second includes steam heat loss in the BTU measurement. This is of importance when measuring the efficiency of how we are handling and transforming energy from the source to end use. I mention this for it is one of the key issues when evaluating the economics and benefits of moving from fossil fuels to renewable energy sources. Thus, much of our homework is already done for us.

The following numbers were generated in order to calculate losses by source and by end use. Likewise, % utilization rates were cross checked in order to estimate 'peak load' splits by category.

The overview shows us that the USA is using 97 quads of energy to produce 31 quads of actual end use or a net loss of 68% during conversion of products, transmission to where it is needed, and actual end use. The electrical side loss is

66%, the transportation side 79%, and the consumer side 60%. In this larger view, we can make three huge and important judgements.

-Using electricity for static use is 13% more efficient than oil for transportation use via fossil fuel engines. Moving from fossil fuel use to run gasoline, diesel, and kerosene engines to renewable electric vehicles should result in significant gains in efficiency in the range of 5%. (From engine to highway.) The overall improvement could be 8% or more.

-Moving the electrical industry from fossil fuels to all renewable energy and battery and other storage devices should result in further gains in the range of 10%. However, if rooftop solar is used, a further 10% savings would be achieved by eliminating transmission, conversion and step-down costs. Thus, rooftop solar for electric car use should be at least 18% more efficient than fossil fuel for vehicles.

-The colossal, and I mean colossal change in this is the change from covering fossil fuel cost to one of zero for renewables. True, it is partly offset by capital cost increases, but with low interest rates, that is a bargain no sane person would pass up!

As for the impact on our balance of payments, it might well be more than 50%. But let's say we have now reduced fossil fuel use by 90%, we still have to explore some other things we can do.

STOP EVERYONE! WE HAVE JUST STATED WE SHOULD MAKE THESE CHANGES AS FAST AS POSSIBLE AND HAS NOTHING TO DO WITH POPULATION OR CLIMATE CHANGE. IN FACT, THAT IS JUST A SIDE BENEFIT, NOT TO MENTION WE HAVE AN EFFICIENT ADMINISTRATION IN PLACE TO MANAGE IT!

Earth, a Planet in Peril

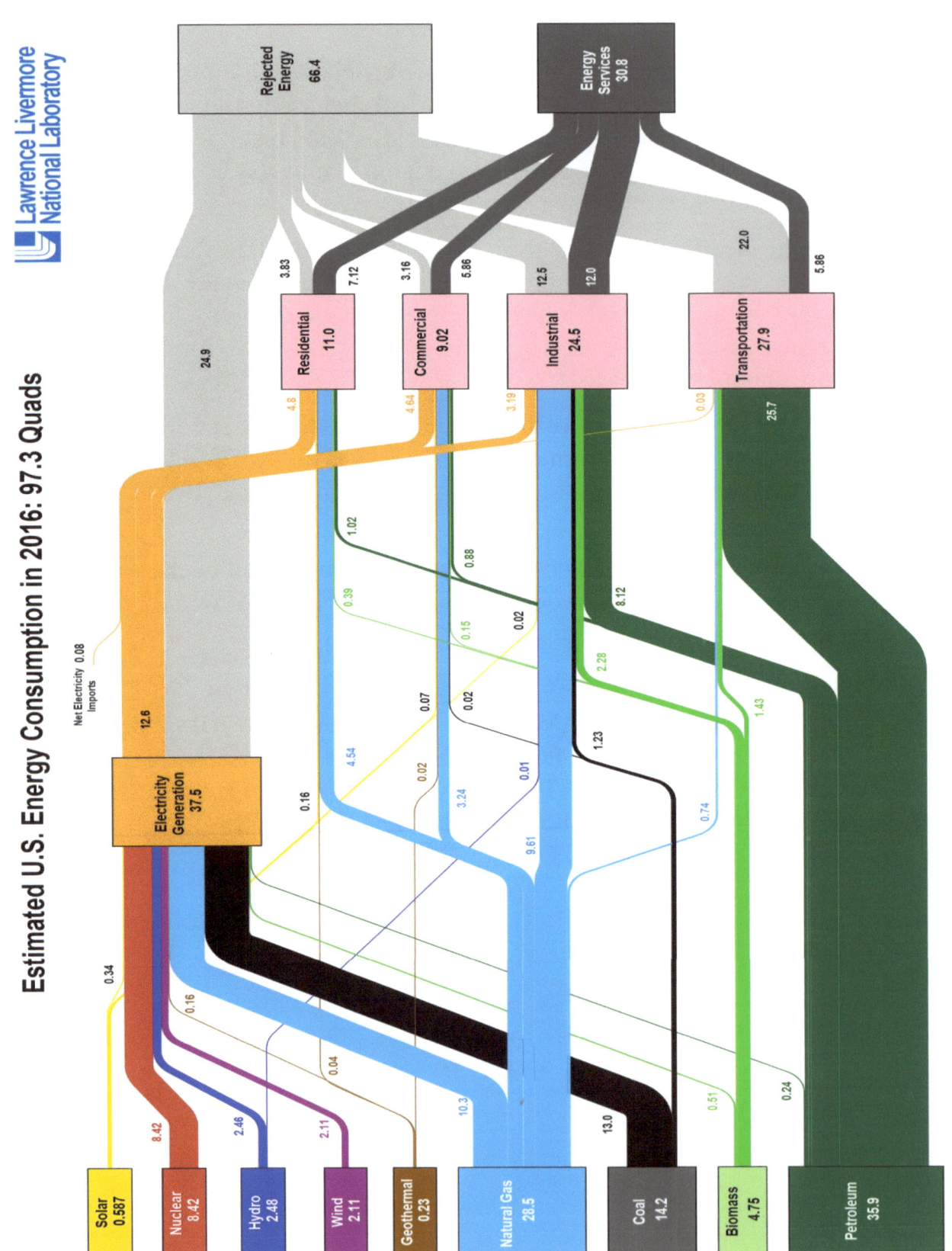

Chart courtesy of LLNL and DOE. (All Data in HHV)

The above numbers were generated in order to calculate losses by source and by end use. Likewise, % utilization rates were cross checked in order to estimate 'peak load' splits by category.

2016 USA Electricity Production

(EIA Thousand Megawatt Hours in LHV).

Source	Production	% Utilization
Solar	56,221	12-30
Nuclear	805,327	75-95
Hydroelectric	265,829	30-65
Wind	226,872	22-65
Geothermal	17,417	68-75
Natural Gas	1,393,295	35-65
Coal	1,240,108	55-75
Biomass	62,572	22-65
Petroleum	23,907	35-65
Total*	4,098,546	

*Includes EIA estimate of 19,467 for small rooftop production not included in data.

*****

9. THE TRANSITION TO RENEWABLES.

Converting the Transportation Industry.

The transportation system involves road, rail, air, and sea travel, and how much of it will have to remain on fossil sources of energy, given today's technology. After that comes the assessment of how renewable energy could be supplied to them and to where. Factoring into all this was the question of whether or not we should encourage the construction of a high-speed rail system such as we find in China and Europe, all to replace much of our inter-city air travel. With the advent of automated driving of cars and trucks and the talk in turn of changing the industry from the ownership of vehicles to a 'ride-hailing' system for commuting and recreational driving, it led to the need to develop some projections that ended up eliminating growth from the equation, i.e. a flat demand case.

The biggest challenge is how these steps can be financed given the fact that our Nation already has Debts totaling $20 trillion and Trade deficits running at $0.7 trillion per year, making the option of Government financing almost a non-starter. With incomes still flat, suddenly adding taxes or fossil fuel energy penalties, to pay for these changes, would be difficult. So, what was the practical solution? To put in place a series of graduated tax penalties and incentives that would give buyers and sellers adequate warning of what was coming such that efficiencies and savings to consumers would offset each other. For example, in the case of automobiles and trucks, imposing a graduated yearly tax on gasoline and diesel fuels, with an incentive to use electric or hybrid vehicles, would alert auto companies to downsize vehicles and improve mileage to offset those costs. Similarly, buyers would also be alerted to make purchase or lease decisions that they could afford. Unfortunately, that calculation has to factor in a reducing fossil fuel use each year. Settling on an escalating tax of 20 cents per year, after 16 years it will have raised $5 trillion, and over twenty years $7 trillion. The good news is that if all of that effort and investment is sourced in the USA rather than imported, then the Nation's GNP will increase resulting in large increases in tax collections and Government budget surpluses. (Assuming our politicians don't mess that up.)

Make no mistake; the technology race is on for electricity storage at several levels. Both encouraging and yet frustrating is the way so many options and uses by others have been blocked by patents, all held in the hope they can generate money for the holders, but at the same time block that idea from general usage. Perhaps a signal that the time for a 'Manhattan Type Project' has come again, with progress and gains to be shared by all? What is happening today is both a slow and expensive way of finding an answer, neither of which we have in plenty. As an example, the most efficient and lowest cost battery storage systems are in the range of $150 kWh for utility companies and $300 kWh for home storage units.

Optimizing the Land Transportation Industry.

Our timing might just be superb, for we are entering a period that will match that of the eighteenth century for invention and innovation. Automation is about to come into its own with the only stumbling blocks being our infrastructure, transportation system, and our imagination. The decisions we make today may well dictate what happens for many years into the future. Let's take one example, the issue of which high-speed land travel system might be best. But first, let's make some ground rules. Whatever system we use should be:

-Free of carbon dioxide and other harmful product emissions.

-Make as little noise as possible.

-Affordable and have a low operating and maintenance cost.

-Efficient with faster transit times.

-Help solve traffic congestion.

-Relatively easy to build and flexible in design and layout.

-Be 100 % American made.

-Quick and easy to gain regulatory and planning approval.

The three options are land, air, and sea.

Land Transportation.

-High-speed rail fails due to the cost of installation (over $100 million per mile or $20 trillion for a 20,000-mile system) and takes years to gain approval. Rail of this type is also inflexible with change both costly and slow. (By comparison, mid-speed trains are cost effective.)

-High-speed road. It might be tempting to discard this option simply by no one having heard of it. But consider this:

-We are in the process of changing over the car, truck and bus industries to becoming all electric (or at least Tesla and the rest of the World is).

-We are also in the process of introducing automated driving.

-The above will include the ability of single units to join with others to form either shared drives or to form trains of interconnected vehicles. The above may well bring back 'The Yellow Cab' era of design when all vehicles are the same rather than being built for individual tastes. (The fuel efficiency savings from forming 'trains' can be up to 25%).

-The above will also include third parties owning or leasing vehicles.

-These vehicles will run on existing but upgraded, highways.

-The advent of these electrically driven automated vehicles traveling in trains with slipstream savings, and with continuous rather than start and stop travel will open up the opportunity to:

a) Build or dedicate highway lanes for high-speed travel.

b) Sort these trains out by destination with the ability of units to join and drop off while in motion.

c) Establish start and stop locations at major highway interchanges to allow train reorganization.

d) Recharging stations would be sited to take advantage of time off for meals or other activities.

e) Vehicles would have to be configured to permit double front and rear connections improved blowout resistant tires and high-speed motors. Similarly, the first vehicle in the train would ideally be a bus with a specially trained driver, recognizing these would be automated.

f) With the above system being low cost, taking advantage of existing infrastructure, offering major reductions in traffic, significant gains in energy efficiency and utilization, plus being flexible, it would seem to be a hands down winner. With interconnected vehicles forming trains for long distance travel, the gain in speed may be such that dedicated lanes for that use may come sooner than we thought. From a technology point of view, 100-150 mph travel should become the norm.

g) Other benefits and socioeconomic changes would also include:

- Reductions in insurance, operating, and in some cases, the cost of owning a vehicle.

- 'Ride-hailing' and shared automated travel dramatically reducing the time for loading and unloading passengers, the need for street parking space and parking space in general.

- As automated truck delivery takes place, the combining of multiple sources and destinations into one truck will occur with a dramatic decline in city truck traffic.

- Likewise, bus traffic and size will be matched to passenger volume.

- Major changes in city street layouts, to optimize the above changes, will result in time and efficiency benefits to all citizens and commuters.

- With more free time for both drivers and passengers, that can be redirected to both work and entertainment, not to mention social activities, the combination of which in daily life could be profound. Who could not be thrilled with these developments and the promised improvements they have for our lifestyle? Again, we left the best to last.

- Imagine, if you will, a city with half of all traffic gone, completely free of fumes and noise, with more space for people, trees, and grassland, how much better would that be for the people, not to mention our planet!

Air Transport.

Air fails for three reasons. No replacement has been found as yet for fossil fuel. Second, air is noisy, expensive, and, when we add in travel to and from airports, a crumbling airport infrastructure, and security issues, a relatively expensive and time-consuming way of travel. The nail in the coffin for that industry is the fact that all near term dreams of cheap and low-cost pollution free air travel hinge on the discovery of ultra-light electric battery technology, something that has yet to be discovered.

The decision by most travelers to fly or drive to their destination is based on travel time and convenience. If one had to make a trip of four hundred miles and had two choices of how to get there, with the first being one's automatically driven car taking three hours door to door from your home to your destination. The second is of course by air involving driving to your local airport, buying a ticket, checking in, perhaps involving checking baggage, waiting for the plane's departure time, hopefully not delayed, then loading, and finally the flight with seemingly ever decreasing service. Once at the other end, the whole process is reversed with baggage waiting a nightmare. Then comes the cab ride or car rental to your final destination. Total time six hours, if you are lucky. Who on earth would make the second choice? In fact, today, 95% of people choose to drive for a four hundred mile or less journey with only 3% by air[23].

Let's take a look at a 750-mile journey. That travel time for automated vehicles increases from three to six hours, whereas the air travel time only increases from six to seven hours. Today, that same study indicates that 35% of travelers chose air before, whereas with these revised times it would change to only 3%. Thus, as the high-speed road becomes more efficient, so will air traffic decline.

The why is simple; time and convenience. The high-speed road will be the game changer for our world. If we doubt that, just think, a seven-hundred-and-fifty-mile radius from any major USA, European or Asian City would cover vast travel centers resulting in a switch of 75% of all air traffic to road. Some examples of what a 750-mile radius covers are:

Washington, DC: Quebec City, Toronto, Chicago, Nashville, Atlanta, Jacksonville.

---

[23] US Dept. of Transport.

San Francisco: Seattle, Salt Lake City, Phoenix, San Diego.

London: all of Europe*.

Paris: All of Europe*.

Rome: All of mainland Europe except Spain*.

Berlin: All of Europe except Madrid.

Istanbul: All of Eastern Europe, West Russia, and the Middle East.

Shanghai: All of China.

Delhi: Karachi, Mumbai, Hyderabad, Calcutta.

*Via 'The Chunnel'

We could expand the list, but the impact on air travel around the world would be extraordinary. Obviously, we need to retain the long-range portion of that industry along with carbon fuels, but with stricter emission standards, which incidentally have been ignored as of this writing.

One last thought about renewable energy: the cost of fuel is zero, true, depreciation and interest costs will offset some of those reductions, but on an incremental basis, and on a balance of trade basis, the impact would be amazing, not to mention the zero emissions.

Sea Transportation.

The EPA, while I have lambasted them, did, however, get it right. That transportation industry has escaped the efforts to clean up our World and should be forced to modernize. This time, if done correctly, it could give the USA and others a wonderful opportunity to build here at home a new and modern fleet of ships that would be non-polluting, environmentally friendly and crewed by duly qualified merchant marine members. What should these new requirements be?

-Tankers to be double hulled with limits on hold size.

-Dry bulk cargo ships should also be double hulled if transporting dangerous or toxic substances.

-All ships should be fully automated. This requirement to include, location, navigation, maneuvering, steering, propulsion and emission recording.

-All ships to be access protected and armed. (Anti-Hijacking protection.)

-All ships to be manned by registered and duly qualified merchant marine crews with captains, first, and second officers on board.

-All propelling equipment to be fully caged to prevent the destruction of large marine creatures.

-All marine engines to meet EPA emission standards. (Requires switch to LNG from bunker, fuel oil, and diesel.)

-All goods into and out of USA ports to be transported in USA ships that meet the above standards.

The time is long overdue to bring to a halt the practice of allowing ship owners to register and operate substandard ships out of Non-United Nations Member Countries, thus avoiding taxes, regulations, environmental, working, and safety standards. What should be one of the USA's largest industries is today almost non-existent!

Conclusion

Who would have thought that investing in renewables would be more efficient and better financially for the USA and the World? Even more stunning is the conclusion it should be done regardless of the benefits to our environment.

With all this coming at a time when we are at a crossroads politically, the timing for making changes could not be better.

THE CONCLUSION AND PLAN

WE SHOULD CHANGE TO RENEWABLE ENERGY REGARDLESS OF POPULATION GROWTH AND CLIMATE CHANGE.

THE PLAN

THE OBJECTIVE IS TO IMPLEMENT THE CHANGEOVER FROM THE USE OF FOSSIL FUELS TO RENEWABLE ENERGY IN AS SPEEDY, ORDERLY, AND AS COST AFFECTIVE WAY AS POSSIBLE BY 2050.

PROPOSED ACTION STEPS.

1). A Manhattan type project be considered for all renewable energy projects.

2). All production of renewable energy equipment to be manufactured domestically. In the event raw materials are required from other Nations, exchanges or shared manufacturing facilities would be permitted.

3). All progress and the monitoring of worldwide activity would be under the direction of the EPA. Despite their record on regulations, they have some of the most knowledgeable people in the industry and are also the best keepers of data. They would in turn report directly to Congress for approval before enacting new regulations.

4). A contingency plan would be developed to artificially cool or warm the Earth by the use of Sulfur and Carbon Dioxides.

5). All Regulations be replaced with a system of monetary rewards and penalties with any additional funds required being financed by a carbon tax.

6). All investments should be booked as Assets and be financed by long term mortgages with depreciation and not the cost the line item in the Nation's Budget. (The debt would also be accounted for as a separate line item from the National debt.)

*****

BIOGRAPHY

My name is Fred Graham-Yooll, born and raised in Scotland, am today a US citizen and make my home in Derwood, Maryland just outside Washington, DC.

Educated in Scotland, Edinburgh Academy, Clifton Hall, and Fettes College. In England, Nottingham University, BSc. Agriculture, Officer in the Royal Artillery. In Canada, joined Northwest Nitro Chemicals, Imperial Oil Ltd, then transferred to the USA with Exxon Mobil Inc. and Esso Malaysia Berhad. In Canada again, as President and CEO of Glacier Ammonia Ltd. Not a believer in retirement, my current enjoyment is writing. 'Cleopatra's Lost Treasure,' Joseph, Moses, and the Exodus, Bumbling Fumbling Stumbling in the Dark, and of course, this book.

Earth, a Planet in Peril

www.ingramcontent.com/pod-product-compliance
Lightning Source LLC
Chambersburg PA
CBHW051150220526
45473CB00003B/717